Synthetic and Degradative Processes in Marine Macrophytes

Synthetic and Degradative Processes in Marine Macrophytes

Proceedings of a Conference held at
Bamfield Marine Station
Bamfield, Vancouver Island
British Columbia
May 16–18, 1980

Editor
L. M. Srivastava

Walter de Gruyter · Berlin · New York 1982

Editor

Lalit M. Srivastava
Department of Biological Sciences
Simon Fraser University,
Burnaby, B. C., V5A 1S6
Canada

CIP-Kurztitelaufnahme der Deutschen Bibliothek

Synthetic and degradative processes in marine macrophytes :
proceedings of a conference, held at Bamfield Marine Station,
Bamfield, Vancouver Island, British Columbia, May 16–18, 1980 /
ed. L. M. Srivastava. – Berlin ; New York : de Gruyter, 1982
ISBN 3-11-008490-2
NE: Srivastava, Lalit M. [Hrsg.]; Bamfield Marine Station

Library of Congress Cataloging in Publication Data

Main entry under title:
Synthetic and degradative processes in marine macrophytes.
Includes bibliographical references and index.
1. Marine algae–Congresses.
2. Marine algae culture–Congresses.
3. Kelp–Congresses.
4. Marine algae–British Columbia–Congresses.
I. Srivastava, L. M. (Lalit Mohan), 1932-
QK570.2.S96 639 81-19513
ISBN 3-11-008490-2 AACR2

 Printing: Kupijai & Prochnow, Berlin. – Binding: Dieter Mikolai, Berlin. Printed in Germany.

CONTENTS

Part III. Marine macrophytes in coastal ecosystems

Part IV. Polysaccharides, kelp farming and harvest

LIST OF CONTRIBUTORS

Asterisk indicates author of a paper in this volume.

*L. ALBRIGHT, Department of Biological Sciences, Simon Fraser University, Burnaby, B.C. V5A 1S6 CANADA

*R.G.S. BIDWELL, Wallace, R.R. #1, N.S. B0K 1Y0 CANADA

E. CABOT, Bamfield Marine Station, Bamfield, B.C. V0R 1B0

*L.M. COON, Marine Resources Branch, Parliament Building, Victoria, B.C. V8V 1X5 CANADA

*L. DRUEHL, Deptartment of Biological Sciences, Simon Fraser University, Burnaby, B.C. V5A 1S6 CANADA

M.J. DUNCAN, Botany Department, University of British Columbia, Vancouver, B.C. V6T 1W5 CANADA

*J.Y. FLOC'H, Laboratoire de Physiologie Végétalé, Faculté des Sciences, Université de Bretagne Occidentale, F-29283-BREST. Cédex, FRANCE

R. FOREMAN, Bamfield Marine Station, Bamfield, B.C. V0R 1B0 CANADA

*M.S. FOSTER, Moss Landing Marine Laboratory, P.O. Box 223, Moss Landing, CA 95039 USA

*V.A. GERARD, Caltech Kerchkoff Marine Laboratory, 101 Dahlia St., Corona Del Mar, CA 92625 USA

*P.J. HARRISON, Department of Oceanography, University of British Columbia, Vancouver, B.C. V6T 1W5 CANADA

C.S. LOBBAN, Biology Department, University of New Brunswick, P.O. Box 5050, St. John, N.B. E2L 4L5 CANADA

*K. LÜNING, Biologische Anstalt Helgoland, Zentrale, Palmaille 9, 2 Hamburg 50 WEST GERMANY

*K.H. MANN, Bedford Institute of Oceanography, Box 1006, Dartmouth, N.S. B2Y 4A2 CANADA

T. McCONNAUGHEY, Department of Oceanography, University of Washington, Seattle, WA 98195 USA

*J. McLACHLAN, NRC Atlantic Regional Laboratory, 1411 Oxford St., Halifax, N.S. B3H 3Z1 CANADA

T.F. MUMFORD, Jr., Division of Marine Land Management, Department of Natural Resources, Olympia, WA 98504 USA

*W.J. NORTH, Caltech Kerchkoff Marine Laboratory, 101 Dahlia St., Corona Del Mar, CA 92625 USA

*J. RAMUS, Duke University Marine Laboratory, Pivers Island, Beaufort, N.C. 28516 USA

R. SCAGEL, Botany Department, University of British Columbia, Vancouver, B.C. V6T 1W5 CANADA

*K. SCHMITZ, Botanisches Institut der Universität Köln, Gyrhofstrasse 15, 5 Köln 41 (Lindenthal) WEST GERMANY

*J. SIBERT, Department of Fisheries and Oceans, Biological Research Station, Nanaimo, B.C. V9R 5K6 CANADA

L.M. SRIVASTAVA, Department of Biological Sciences, Simon Fraser University, Burnaby, B.C. V5A 1S6 CANADA

R. WAALAND, Botany Department, University of Washington, Seattle, WA 98195 USA

*W.N. WHEELER, Department of Biological Sciences, Simon Fraser University, Burnaby, B.C. V5A 1S6 CANADA

J.N.C. WHYTE, Fisheries and Marine Service, 6640 N.W. Marine Drive, Vancouver, B.C. V6T 1S6 CANADA

*J. WILLENBRINK, Botanisches Institut der Universität Köln, Gyrhofstrasse 15, 5 Köln 41 (Lindenthal) WEST GERMANY

OTHER PARTICIPANTS

C. Bell, R. Bocking, R. Boal, M. Bonettemaker, R. Bunting, B. Carowsell, M. Fabijan, I.M. German, J. Lindsay, K. Lloyd, G. Mogensen, B. Pakula, L. Quain, A. Quek, K. Rosell, P. Ryan, B. St. Clair, R. Saunders, M.L. Shih, M. Shivji, R. Smith, S. Smith-Pakula, G. Stalder, A. Stewart, D. Trotter, T. Tuominen, I. Velji, S. Villeneuve

PREFACE

Seaweeds are important sources of food, several phycocolloids, pharmaceuticals and, more recently, biomass for energy. Their continued and successful commercial utilization depends on our understanding of their growth and reproduction, their strategies and adaptations that enable them to exploit their environment, and their place in the coastal ecosystem of which they are such an important part. Among the seaweeds, the brown (Phaeophyta) and red (Rhodophyta) algae are specially important, not because they are highly visible, but because they form the basis for much of the seaweed industry. Despite their importance, however, these macrophytes, which include the "redwoods of the sea," are poorly studied and very few volumes have been exclusively devoted to them. The present volume is a record of the proceedings of a small conference held at the Bamfield Marine Station, Vancouver Island, British Columbia on May 16 - 18, 1980 on the synthetic and degradative aspects of marine macrophytes.

Over the years, a small but active group of researchers studying various aspects of macrophytes, especially kelp growth and physiology, reproduction, chemical constituents, mariculture and degradation, has come together in British Columbia and it seemed appropriate that the Bamfield Marine Station, where much of this work is being done, be the site of this first conference. Since the board and residence facilities at Bamfield were very limited, only a small number of participants could be invited. While this resulted in a narrowed scope for the conference, it permitted a greater exchange of ideas on the topics that were covered than is usually possible in larger conferences. Also, the unhurried contact of people from different disciplines under the idyllic setting of a fishing village on the Pacific ocean, far from the madding crowd, proved to be a most valuable experience.

The aim of the conference, set by the Organizing Committee, was to emphasize marine macrophytes as a natural resource and their prudent management. Since kelp are the predominant macrophytes along the Pacific coast of North America, they received the most attention. A mixture of topics on growth and reproduction, nutrient availability and supply, macroalgal associations and degradation, and on kelp farming and harvest, was proposed and within that framework, individual speakers were left free to develop their themes and highlight what they thought was important. During the conference it was agreed by the participants that the papers and relevant discussions, which were being taped, be published as a single volume. On returning to their respective homes, the authors prepared their manuscripts according to the publisher's instructions and sent them back to me, in most cases promptly and without any nudging. In two cases, however, I am still awaiting the manuscripts.

It has been my very pleasant duty to compile these proceedings. The papers, all of them by experts in their fields, are of consistently high quality and, having spent the better part of my research career with higher plants, I am impressed by the range of unique and in some cases untouched problems that marine macrophytes offer for inquisitive, bright, and resourceful young minds. At times I have changed the order and grouping of papers from that followed in the conference for better continuity and readibility. I would like to take this opportunity to thank all contributors, authors and discussants, who made this volume possible, and who so patiently and ungrudgingly bore the brunt of my editorial scissors.

Many individuals and institutions cooperated and helped in the production of this book. Financial support for the conference was provided in part by Simon Fraser University, in part by the Marine Resources Branch, Department of Environment, Government of British Columbia, and in part by the participants who not only generously donated their time, but also bore a substantial part of their travel fare to Vancouver and back. Our special thanks are due to the Director and staff of Bamfield Marine Station who provided the facilities for holding the conference and to numerous students, staff and faculty who volunteered their time and effort to running the various errands associated with the conference. I would specially like to thank Larry Albright, Mia Bonettemaker, Bea Donald, Louis Druehl, Walter Griba, Lin Kemp, Carolyn Tate, and Carole Thompson, all from Simon Fraser University, and Ron Foreman from University of British Columbia and Bamfield Marine Station.

Funds for preparing this volume for publication were provided by Simon Fraser University. It is not often that the editor of a book is blessed with a typist who is a specialist in that field and who also owns a word processor. I am extremely grateful to Bill Wheeler for being here at the right time to do the typing for this book and to decipher many mumblings on the recording tape. I hasten to add, that while the typographical errors may be Bill's or the word processor's, the editorial mistakes are all mine.

Finally, I would like to thank Walter de Gruyter for publishing this volume and for being very patient with an editor who failed to meet several deadlines.

July 1981 Lalit M. Srivastava

Organizing Committee:

L. Albright, L. Druehl, R. Foreman
C.L. Kemp, L.M. Srivastava

OPENING REMARKS

R.F. Scagel

Department of Botany, University of British Columbia,
Vancouver, B.C. Canada

On behalf of the Management Council of the Western Canadian Universities Marine Biological Society, I am pleased to welcome the participants to this Conference. This society is a consortium of five universities, comprising the University of Alberta, the University of British Columbia, the University of Calgary, Simon Fraser University and the University of Victoria. Although the Society has not given any special mandate to marine plant biologists at the Bamfield Marine Station, it has been highly supportive of their efforts, and is very appreciative of the high profile given to this station by marine plant researchers, especially through the contributions of Drs. Druehl, Foreman and Srivastava.

Although Dr. Srivastava was most circumspect when he invited me to give some opening remarks at this Conference, I detected that the reason was that I might be regarded as a senior marine plant citizen - hopefully not yet a fossil specimen. And when I examined the roster of invited participants, which included former graduate students, post doctoral fellows, individuals whom I had met as graduate students elsewhere, and colleagues whom I had introduced to local kelp beds or known for many years, all of whom are now internationally known for their research on seaweeds, it seemed that I had little choice but to accept the label in the face of this evidence. Furthermore, it is 34 years ago this month that, as a graduate student, I set forth in a 9-foot rowboat to spend the summer studying a bed of Macrocystis near Port Hardy, British Columbia (1), at the commission of the British Columbia Provincial Fisheries Department, and at a stipend of $150.00 per month. At that time it was this Department that was responsible for licensing seaweed harvesting in British Columbia. Hence, perhaps I can give an historical perspective to this Conference, with some emphasis on the local scene. What is the present state of fundamental knowledge concerning seaweeds? What are the prospects for application of the great wealth of expertise that is outlined in the programme of this conference? Wisdom doesn't necessarily come with age, but perhaps one can learn something from past mistakes!

Early in the 20th Century the United States Government became concerned about the extent to which it relied on imported sources of potash for agricultural fertilizer. This led to a massive report (2) published by the United States Department of Agriculture in 1912 on fertilizer resources, including kelps as a source of potash, and extensive surveys of these marine plant resources of the Pacific Coast of the United States, including Alaska, were undertaken. Shortly thereafter, in 1916, the Fisheries Research Board of Canada (3) carried out a survey of the kelp beds along the coast of British Columbia, again with a view to possible utilization of kelp resources as a source of potash. During World War I years (1914-1918), when the source of potash from the German mines at Stassfurt was cut off, the United States produced some potash from kelp (e.g., in 1917, 3,572 tons of potash were produced). However, shortly thereafter extensive deposits in Chile became available to world commerce and it was no longer economically feasible to use seaweed resources as a source of fertilizer potash.

Although interest in kelp as a commercial source of potash disappeared, an even greater interest in kelps as source of phycocolloids soon emerged and the foundation was laid for the modern algin industry in California based on Macrocystis. An increasing demand for algin, and the shortage of supplies of agar from Japan, resulting from World War II (1939-1944), again gave an impetus to the seaweed industry on the Pacific Coast of North America. However, despite the availability of exploitable natural kelp resources in British Columbia, attempts to develop a seaweed industry of any magnitude have still not met with success.

In the mid-forties, there was renewed interest on the British Columbia coast in seaweeds, and some funds were provided for surveys and research by the British Columbia Provincial Government (through the Provincial Department of Fisheries and the British Columbia Research Council), the Canadian Federal Government (through the Fisheries Research Board), and a private industry in Maine that was interested in agar. A kelp company was finally established, tried to get into operation in a plant constructed on Deer Island in the vicinity of Hardy Bay, and almost succeeded by 1947. From the beginning, it was planned that this company would produce algin, although there were also plans to produce seaweed meal, pills, and possibly even fertilizer. Apart from producing some pills and a small amount of meal, this company never achieved full production and soon went into receivership at a loss of several hundred thousand dollars. Despite the availability of sufficient kelp (mainly Macrocystis) to have maintained a modest operation and sufficient private capital to have anticipated success, the company failed apparently because of poor management.

In the mid-sixties there was another attempt to establish an algin industry in British Columbia - this time at the north end of the Queen Charlotte Islands. For reasons of apparent poor management and financial problems this effort also soon failed, at a considerably greater financial loss than the earlier attempt in the mid-forties.

There has been a renewed interest and support of the British Columbia Provincial Government in both fundamental and applied research on marine plants in the past decade. This renewed interest on both kelps as well as on other phycocolloid-producing marine algae, if matched with financial backing and entrepreneurial skill, may lead to the successful establishment of a healthy seaweed industry in British Columbia. Marine plant scientists in British Columia face a special challenge in this development.

The research activities to be reported on and the presence of many distinguished scientists as participants in this Conference are bound to have a significant impact on local interests and will undoubtedly have a positive influence on the development of a healthy seaweed industry in British Columbia. Whatever small role I have played as a pioneer in encouraging and developing marine phycology in British Columbia is more than amply rewarded in finding an assemblage of expertise such as is represented here today meeting in British Columbia and at the Bamfield Marine Station.

References

1. Scagel, R.F.: British Columbia Dept. Fish. No. 1, pp. 1-70 (1948).

2. Cameron, F.K.: U.S. Senate, Doc. No. 190, 62nd Congr., 2nd Sess., pp. 1-290 (1912).

3. Cameron, A.T.: The commercial value of the kelp-beds of the Canadian Pacific coast. A. Preliminary report and survey of the beds. Contr. Canadian Biol. for 1914-15, pp. 25-39 (1916).

Part I. Strategies of growth and reproduction

PHYSIOLOGY OF CARBON FIXATION IN BENTHIC MARINE ALGAE

J. Willenbrink

Botanisches Institut, Universität Köln
D-5000 Köln-Lindenthal, F.R. Germany

Life would not be possible without the light energy of the sun. The process of capturing and utilizing this energy has been perfected over millions of years of plant evolution. In photosynthesis, light energy is converted in a series of steps into chemical energy which can then be used by the organism for various metabolic processes. The typical photoautotrophic growth of algae includes both this energy conversion and carbon fixation, also called "dark reaction," in which the chemical energy in the form of ATP and $NADPH_2$ is used to bring CO_2 into reduced organic compounds.

I would like to concentrate here on some aspects of the physiology of CO_2 -fixation rather than on its biochemistry. The first section will be devoted to autotrophic CO_2 -fixation, the second to location of the photosynthetic carbon reduction cycle in the algal cell, the third to CO_2 -fixation by systems other than RuBP-carboxylation, and the fourth and last to regulation of CO_2 -fixation.

Autotrophic CO_2 -Fixation

Marine algae are exposed to carbonic acid (H_2CO_3) and its ions, bicarbonate (HCO_3^-) and carbonate (CO_3^{2-}) as well as to CO_2, the last one at a very limited concentration. In addition, an increase in pH results in an increase in the amount of bicarbonate and carbonate in the sea water.

These algae are able to take up CO_2 which readily diffuses across cell membranes (25) and can be directly used for carboxylation by RuBP-carboxylase. Many species also use bicarbonate but this must be

Synthetic and Degradative Processes in Marine Macrophytes

converted to CO_2 before it is assimilated. Such a conversion may occur in either of two ways: direct uptake of bicarbonate by the cells and subsequent hydrolysis to CO_2 by carbonic anhydrase (25) a process which might not be present in all algae (11); second, bicarbonate may be converted to CO_2 before it has been taken up by the plant. In sea water, bicarbonate concentration is more than 100-fold that of CO_2, and hence, the latter process is quite likely. Both carbon sources may play a role in the same species, for as Kremer and Schmitz (15) have shown recently the littoral species of *Fucus platycarpus* may have the same rate of CO_2 -fixation in submerged as well as in emerged state.

The important steps in incorporation of CO_2 into organic compounds and their subsequent reduction, the "dark reaction," have been elucidated by Calvin and his group. By 1969 the existence of the reductive pentosephosphate cycle, the Calvin cycle, was established for numerous chlorophytes, but there was very little published information on CO_2 -fixation in other algae, especially marine algae. The pioneering work by Yamaguchi *et al.* (35) on CO_2 -fixation by brown algae was a step forward in this direction.

CO_2 -fixation in brown algae

Yamaguchi *et al.* (35) incubated fronds of the Pacific brown alga, *Eisenia bicyclis*, with ^{14}C-bicarbonate sea water in the light and followed the intramolecular distribution of label in mannitol, which is known to be the main assimilate of these algae, in 1 to 60 min periods. The ratio of total ^{14}C to ^{14}C in C_1 and C_6 positions in the mannitol molecule decreased from 8.0 after 1 min to about 3 after 60 min, which suggested a trend toward a uniform distribution of the label. These results were subsequently confirmed in our laboratory using *Fucus serratus* and *Dictyota dichotoma*. In both species, the ratio of total ^{14}C to ^{14}C in C_1 and C_6 positions decreased in experiments from 6 s to 600 s ^{14}C assimilation (32). These results imply that mannitol may be synthesized *via* the reductive pentosephosphate cycle through 3-carbon compounds. Yamaguchi already had demonstrated enzyme activities in the crude extract of *Dictyota dichotoma* for hexose-diphosphatase, mannitol-1-phosphatase, and glucose-phosphate-isomerase, which can be involved in mannitol biosynthesis. Because of the high mucilage content and the presence of large amounts of phenolic compounds, progress in enzyme work on brown algae was not really possible until technical improvements in the isolation of enzymes could be made, for instance, by addition of polyvinyl pyrrolidone (PVP), bovine serum albumin (BSA) and cysteine. Nordhorn et al. (23) demonstrated the presence of both glyceraldehydephosphate-dehydrogenase and mannitol-1-phosphate-dehydrogenase in Fucus serratus, both of which are involved in mannitol biosynthesis. In 1973, evidence was presented for RuBP-carboxylase in Laminaria, which catalyzes CO_2 binding in the light (29).

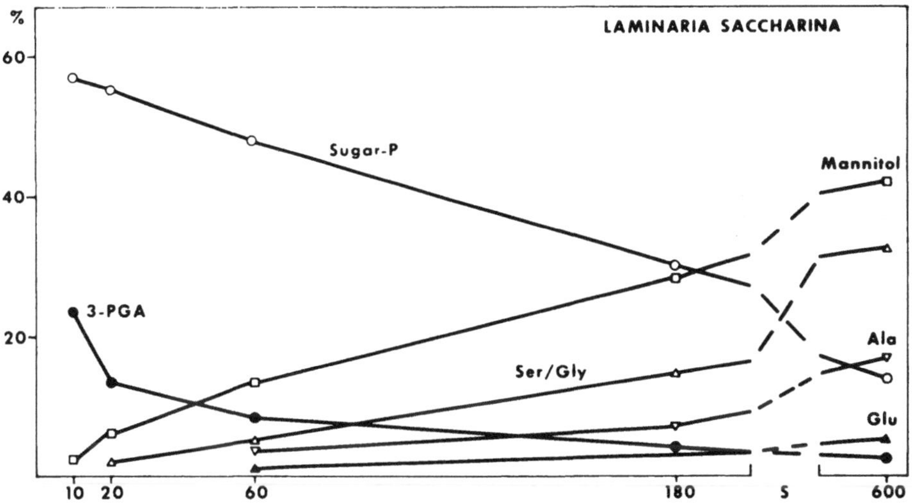

Fig. 1: Labelling pattern of assimilation products of the blade of Laminaria saccharina. Data are percentage of total ethanol-water-soluble radiocarbon, present after 10 - 600 s of photosynthesis (after 17).

In other work on CO_2 -fixation by brown algae done in our laboratory, it has been established that the intermediates of CO_2 -fixation and assimilates show clear kinetics of the distribution of ^{14}C with increasing time of exposure of the frond to $^{14}CO_2$ (12,17). Figure 1 presents data on short-term experiments with Laminaria saccharina. The rapid decrease of relative label in PGA corresponds to the well established fact of PGA as being the first product of carboxylation in the light. The label in sugar-phosphates also decreases with time, as expected, whereas mannitol, serine/glycine and alanine become heavily labelled with time, both in relative and absolute terms. It should be noted that the total percentage of label in the four amino acids listed here exceeds that in mannitol and accounts for more than 50% of the total radioactivity present in the frond after 10 min. Similar experiments were carried out with some other brown algae which all showed a similar pattern of labelling (e.g. Giffordia mitchellae and Laminaria hyperborea), except that in Fucus the increase in ^{14}C-labelled amino acids was slightly less (14). These data correspond well with the results of Hellebust and Haug (9) who found amino acids, especially alanine and aspartate, more strongly labelled than mannitol in 3 hour incubation experiments with Laminaria digitata.

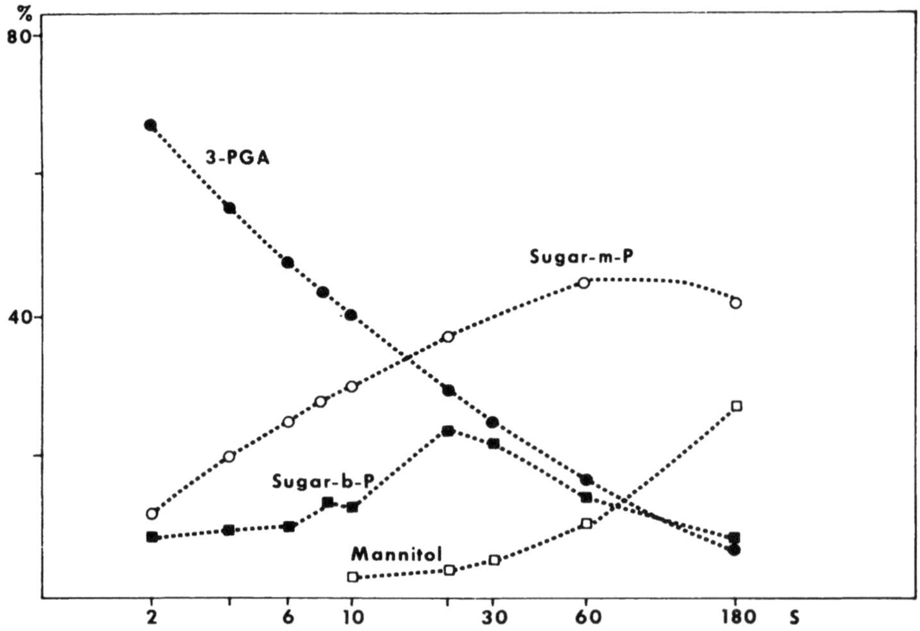

Fig. 2: Time-course of ^{14}C in various phosphate esters and in mannitol after 2 to 180 s ^{14}C -photosynthesis by blade areas of Laminaria hyperborea. Blades were pre-illuminated for at least 10 min prior to exposure to radiocarbon. Data are percentage of total ethanol-water-soluble radiocarbon (after 33).

For Laminaria hyperborea, short-term incubations down to 2 s show an even more pronounced decrease of relative ^{14}C label in PGA (Fig. 2). By separating sugar-biphosphates (SBP) from sugar-monophosphates (SMP), two subsequent peaks could be shown: the peak for SBP might represent mainly RuBP but fructose biphosphate is not excluded, the peak for SMP might contain various sugar-monophosphates. The subsequent rise of ^{14}C-mannitol with a corresponding decline in ^{14}C-labelled PGA and sugar phosphates, indicates that mannitol is one of the main assimilates in brown algae (33).

CO_2 -fixation in marine red and green algae

For Rhodophyceae, Bean and Hassid (3) presented evidence that floridoside became labelled in CO_2 -fixation, and that Calvin cycle was operating in these algae. A study on Porphyra and other genera revealed that floridoside and digeneaside were the most strongly labelled carbohydrates; mannitol could not be detected (16). The pattern of ^{14}C-labelling in the photosynthetic intermediates reflects the role of glycerophosphate as a precursor of floridoside. Glycine and serine, compared with brown algae, carry a lesser but still a remarkably high percentage of total ^{14}C-label. The prevalence of amino acids in rhodophycean photosynthesis becomes even more obvious in long-term experiments. In Pterosiphonia dendroidea, 75% of the total assimilated carbon was distributed in the amino acid fraction (13), a fact which emphasizes the extraordinarily high rate of amino acid synthesis in photosynthetic carbon fixation in many marine algae.

Information on CO_2 -fixation in marine chlorophytes is relatively meagre. Percival (24) described that in Ulva lactuca sucrose was the most highly labelled sugar, with some ^{14}C-label in glucose, fructose, xylose, and myoinositol, in experiments longer than 20 min. For the siphonaceous green alga, Caulerpa simpliuscula, Howard and coworkers (10) have shown soluble β-1,3-polyglucans as important labelled carbohydrates, whereas sucrose, as might have been expected, plays only a minor role. The authors presented data which show an exceptionally high label in sugar phosphates, especially phosphoglyceric acid and fructosebiphosphate, with incubation time of up to 40 min, and suggest a role for these substances as a storage pool. That speculation may be questionable, but the other data presented by these authors show a relatively slow synthesis of carbohydrates, and, hence, the high pool of phosphate esters appears reasonable. Again, the radioactivity of amino acids increased markedly, alanine became labelled most rapidly, followed by glycine and others, but unfortunately the data given by the authors do not allow a calculation of percentage of total ^{14}C in the amino acid fraction.

Site of Photosynthetic Carbon Reduction in Algae

From the points of view of energetics as well as our understanding of the compartmentation inside a photosynthesizing cell, it is important to know whether the chloroplasts are the sites of the Calvin cycle, and further, where in the cell the synthesis of the final assimilate takes place. There is good evidence in the literature from higher plants that the enzymes of the Calvin cycle are exclusively located in the chloroplasts, whereas the enzymes involved in sucrose biosynthesis may be active outside these organelles (5). Bidwell _et al._ (4) reported on preparations of chloroplasts from _Acetabularia mediterranea_ which carried out photosynthetic activity _in vitro_ that was indistinguishable in rate and products from photosynthetic activity in intact cells. Since these preparations were able to synthesize sequentially oligosaccharides from sucrose and were shown to be relatively clean, the chloroplast of _Acetabularia_ has been suggested as the site of synthesis of sucrose. However, these preparations still contained cytoplasm, as was shown later (34).

From the above review, it might be concluded that sucrose is synthesized adjacent to but outside the chloroplasts. In order to test whether or not chloroplasts in brown algae (phaeoplasts) are the site of mannitol synthesis, we tried to isolate phaeoplasts after _in vivo_ exposure of fronds of _Laminaria hyperborea_ and _Fucus serratus_ to $^{14}CO_2$ for various times. It was impossible to isolate phaeoplasts from _L. hyperborea_ fronds, but with _F. serratus_ there was partial success using a nonaqueous procedure (30). After extraction and chromatography, distribution of the label in the individual photoassimilates and total ^{14}C between phaeoplasts and the remainder of the cell was analyzed. After 10 s ^{14}C -assimilation, more than 80% of all ^{14}C-mannitol was present within the plastid fraction, but after 180 s the major part was found in the remainder of the cell

(Fig. 3). These data support the idea that mannitol is synthesized inside the phaeoplast.

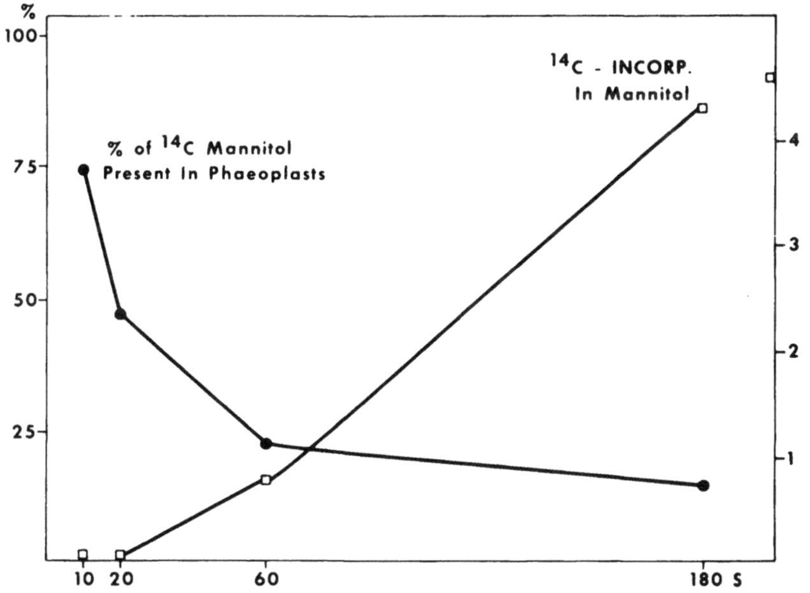

Fig. 3: Time-course of ^{14}C -incorporation into mannitol present in phaeoplasts isolated nonaqueously from fronds of Fucus serratus which were exposed to $H^{14}CO_3^-$ - seawater for 10 - 180 s in the light. Note the high percentage of labelled mannitol present in the phaeoplast fraction after 10 s exposure of the frond to radioactive bicarbonate, and its subsequent and rapid decline (modified after 30).

In a second approach, we used aqueously isolated phaeoplasts from F. serratus (23). With these preparations, it has not been possible so far to obtain rates and photosynthetic activities comparable to those reached for higher plant cells, but the phaeoplast fraction seems to be able to synthesize mannitol. In addition, O_2 evolution by the phaeoplasts shows light saturation at 58 cal m^{-2} h^{-1}; O_2 evolution is temperature insensitive between 4 C and 24 C, and above 24 C the rate of apparent photosynthesis decreases rapidly. This sensitivity to even a slight rise in temperature above 24 C is typical for benthic brown algae. DCMU, a competent inhibitor of photosystem II, has a half-maximal inhibition at 5 x 10^{-7} M. Finally, the in vivo absorption spectrum exhibits a peak for fucoxanthin which coincides with the action spectrum indicating that fucoxanthin is involved in light reactions (23). The distribution of enzymes involved casts some doubt on mannitol being synthesized in the phaeoplasts. So far, mannitol-P-dehydrogenase has been found mainly outside the phaeoplasts, a contradiction which cannot be resolved as yet. It is noteworthy in this connection that phaeoplasts in brown algae are surrounded by periplastidal cisternae of the ER system (6), which may contain enzymes of hexitol biosynthesis. Presumably, the periplastidic network has not been separated completely from the phaeoplasts during isolation. From an energetic point of view and considering compartmentation, mannitol may well be synthesized outside the phaeoplasts if phosphate exchange across the plastid membrane is linked to triosephosphate transport, as stated by Heber (8).

Other CO_2 -Fixation Reactions

Whereas CO_2 -fixation outlined so far is intimately related to energy conversion, further CO_2 -fixation reactions are important for all growth in autotrophic as well as in heterotrophic organisms. These reactions are well known for animal and higher plant systems, but they have received little attention for algal cells, especially marine algae. For instance, for marine algae there are no data in the literature on the synthesis of arginine and prymidines, dependent on the activity of an enzyme which catalyzes carboxylation of ammonia yielding carbamoyl-phosphate.

Other carboxylating enzymes (excluding the biotin-dependent enzyme which needs CO_2 only as a catalyst to start fatty acid synthesis), are mainly involved in anaplerotic processes which supply carbon skeletons for biosynthesis of various products, and in the absence of which intermediary pathways like glycolysis or the Krebs-cycle would be depleted of their intermediates. An example is phosphoenolpyruvate-carboxylase whose reaction yields oxaloacetate, the acceptor for acetyl-CoA and substrate for aspartate biosynthesis. This enzyme is present in most of the organisms studied. In this connection, it should be noted that PEP-carboxykinase is a highly active enzyme in some marine algae (29).

Table I: Carbon fixation in the light and under dark conditions in young and old parts of blades of Laminaria hyperborea.

	Light fixation µmol C/dm²/h	Dark fixation µmol C/dm²/h
Young frond		
February, 4 C	15	4.1 (27.3)[a]
March, 4 C	26	3.6 (13.8)
April, 5 C	15	4.2 (28.0)
May, 8 C	23	2.9 (12.6)
Old frond		
February, 4 C	23	1.4 (6.1)
March, 4 C	35	1.5 (4.3)
April, 5 C	36	1.5 (4.1)
May, 8 C	37	0.8 (2.2)

[a] Figures in parentheses represent % of carbon fixed in light. Compiled after (33).

This enzyme used to be considered a decarboxylating enzyme, but an increasing number of studies in higher plants indicate that it has a carboxylating function. Akagawa et al. (1,2) presented data on carboxylation rates in some species of Pacific brown algae and identified PEP-carboxykinase as the enzyme responsible. Our studies on carboxylation processes as affected by age of fronds in Laminariales revealed exceptionally high rates of dark fixation of CO_2 in the meristematic zone or young parts of the fronds. In Laminaria hyperborea photosynthetic rates in young and old parts were in the same range, 12 to 40 µmol CO_2 dm^{-2} h^{-1}, at 4 C which is the normal ambient temperature in the North Sea during early spring (Table I). However, dark fixation was much higher in the young than in the old parts of the frond, namely 13 to 28% of the corresponding light fixation in the young frond and 2 to 6% in the old frond (33). Principally the same correlation can be deduced from data on Macrocystis integrifolia, Nereocystis luetkeana and several other brown algae (31). The pathway of ^{14}C-incorporation in the dark is completely different from that in the light, and label from dark fixation can enter phosphate esters or mannitol only by gluconeogenesis (see Fig. 5).

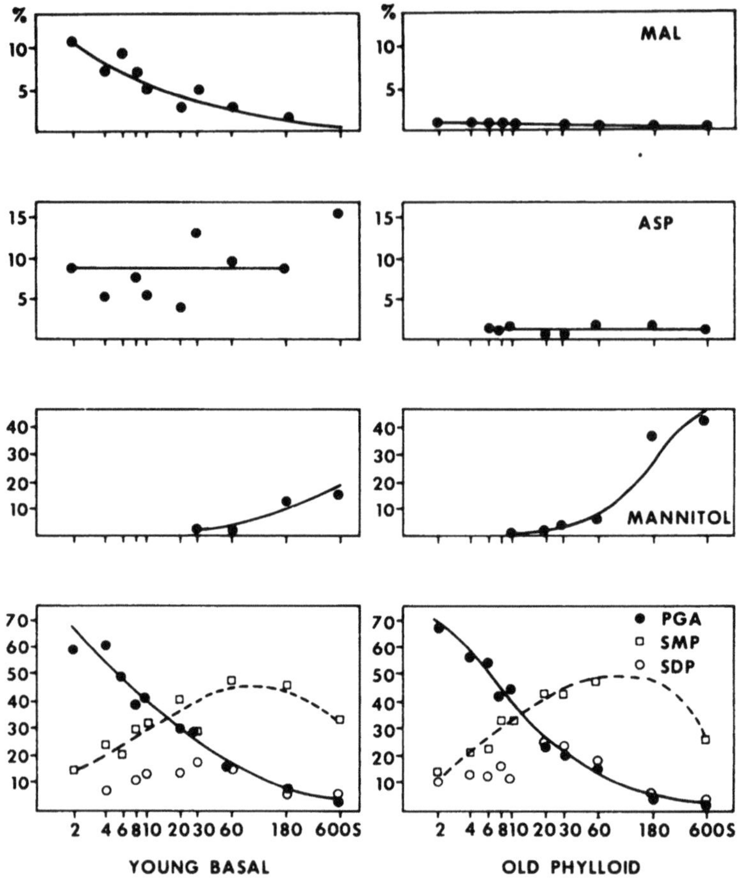

Fig. 4: Kinetics of ^{14}C-incorporation into malate (MAL), aspartate (ASP), mannitol, phosphoglycerate (PGA), and sugar mono- or di-phosphates (SMP or SDP) in young and old parts of blades of Laminaria hyperborea. Samples were exposed to $H^{14}CO_3^-$ - seawater in the light for 2 - 600 s; ambient temperature 4 C, February. Data presented as % of total ethanol-water-soluble ^{14}C (modified after 33).

Short term $^{14}CO_2$-incubation experiments, 2 to 600 s duration, show that in the young frond two types of carboxylation reactions occur in the light (33). A coupling of these carboxylating reactions, as in C-4 plants, is a tempting speculation, but two types of evidence indicate that both systems work independently. The ratio of labelling of the first carboxylation products, phosphoglycerate to malate/aspartate, is 70% to 30% after 2 to 4 s $^{14}CO_2$-incubation of frond segments under continuous light (Fig. 4, left set of drawings). Secondly, the shift of short term labelled intermediates, after a subsequent transfer of the labelled frond samples to a CO_2-free sea water shows a marked increase in ^{14}C-RuBP, but only a marginal increase in ^{14}C-PEP, which suggests that both systems are active in the light but only one of them (RuBP carboxylase) is light dependent. There is no doubt that both systems are involved in metabolism of carbon. Figure 5 summarizes a possible scheme linking the site of action of PEP-carboxykinase to intermediary metabolism and various synthetic pathways in marine algae.

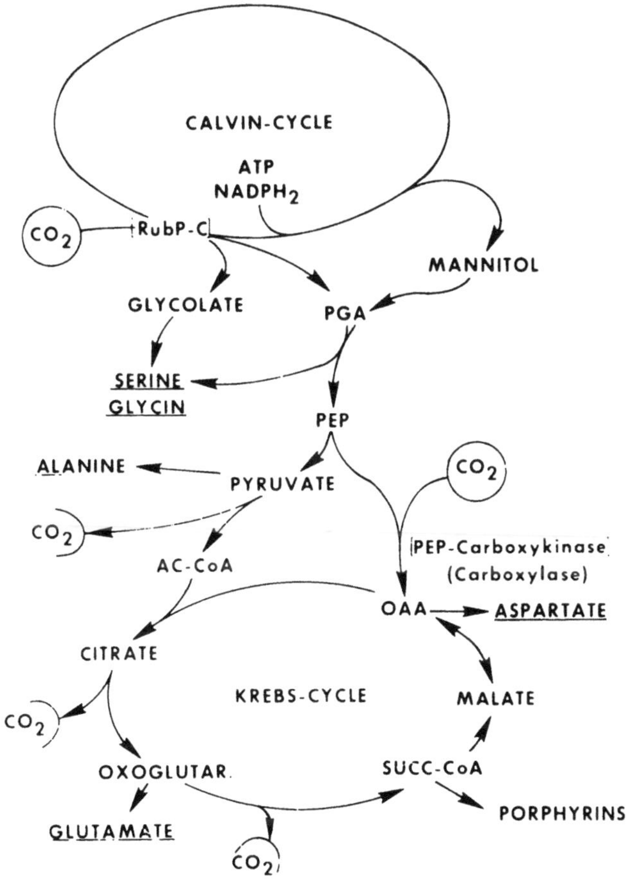

Fig. 5: Flow of carbon between Calvin-cycle and Krebs-cycle in brown algae, with special reference to ribulosebiphosphate-carboxylase and phosphoenolpyruvate-carboxykinase as primary sites of CO_2- fixation. Amino acids are underlined.

Regulation of CO_2 -Fixation in Marine Algae

It is obvious that among external factors light and temperature play the dominant role because for marine organisms inorganic carbon supply and oxygen level seem to be maintained more or less constant throughout the whole year. However, there is some evidence that nutrient supply, mainly of nitrogen, governs growth rate in some species (7). As far as light is concerned, irradiation leads to species-specific enhancement of CO_2 -fixation. For species of Laminaria, Macrocystis, Nereocystis, and Fucus, light saturation is reached at about 200 foot candles (Fig. 6; see reference 31). Consequently, these species behave, in a classical scheme, as shade plants which presumably is of ecological significance. It was pointed out at the beginning, that light influences CO_2 -fixation by providing ATP and $NADPH_2$, whose availability in light provides the striking difference between photosynthesis and dark fixation; for it is not until reduction of carbon has occurred that a real gain is made by the plant. As shown by short-term light/dark experiments, ATP-synthesis and corresponding decrease in ADP occurs with the same velocity in both higher plant and brown algal cells (23).

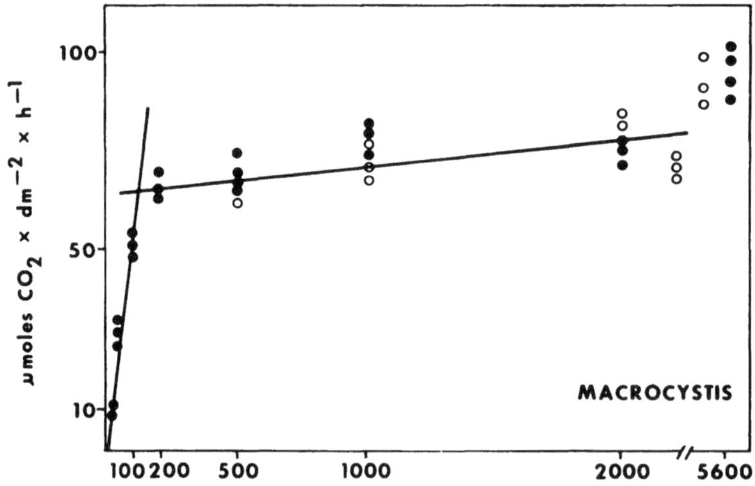

Fig. 6: Dependence of net photosynthesis on incident light (in ft. candles along x axis) in fronds of Macrocystis integrifolia. Results from 2 experiments, open and closed circles, are shown. Assimilation rate measured by ^{14}C- fixation (modified after 31).

It is well known from higher plants that they have evolved a long distance transport system for assimilates, which is of significant advantage in the carbon economy of the organism. Mobilization of food material stored in older parts of the thallus and translocated to young growing tissues, occurs in higher organized brown algae, also. Evidence for such translocation was presented by Lüning for Laminaria hyperborea. Growth of this alga starts during a season when light intensity in the sublittoral zone exceeds the compensation point only for a few hours daily (21). Growth measurements clearly demonstrated a dependence of the young frond on materials translocated from the old one (20). By feeding a distinct part of the old frond with ^{14}C-sea water, it was shown that some hours later radioactive material moved mainly towards the new frond where the label was concentrated in the basal zone (27). Detectable movement occurred only in the growth period of the new frond. In a first, rather rough calculation, it was shown that about 50% of the growth of the young blade was due to translocation of newly synthesized material and the other 50% to translocation of mobilized material from the old frond (22). Translocation is by no means specific for Laminaria hyperborea but has been shown for all Laminariales studied so far (see Schmitz, this volume). Leaving aside the interesting question of the transport process itself, it may be mentioned that these plants seem to translocate what they produce as photoassimilates; in the conducting system, there are both high amounts of amino acids and mannitol (26-28).

The question arises whether or not the high rate of dark fixation in a young frond contributes to its growth. In order to answer that question, it must be remembered that anaplerotic reactions become efficient only when they are provided with reduced materials. As reviewed earlier, the main translocate is mannitol, a highly reduced compound. The establishment of a system which combines light-dependent carboxylating reactions with a supply of reduced material could have been a necessity in the course of evolution considering the biotope of this perennial alga.

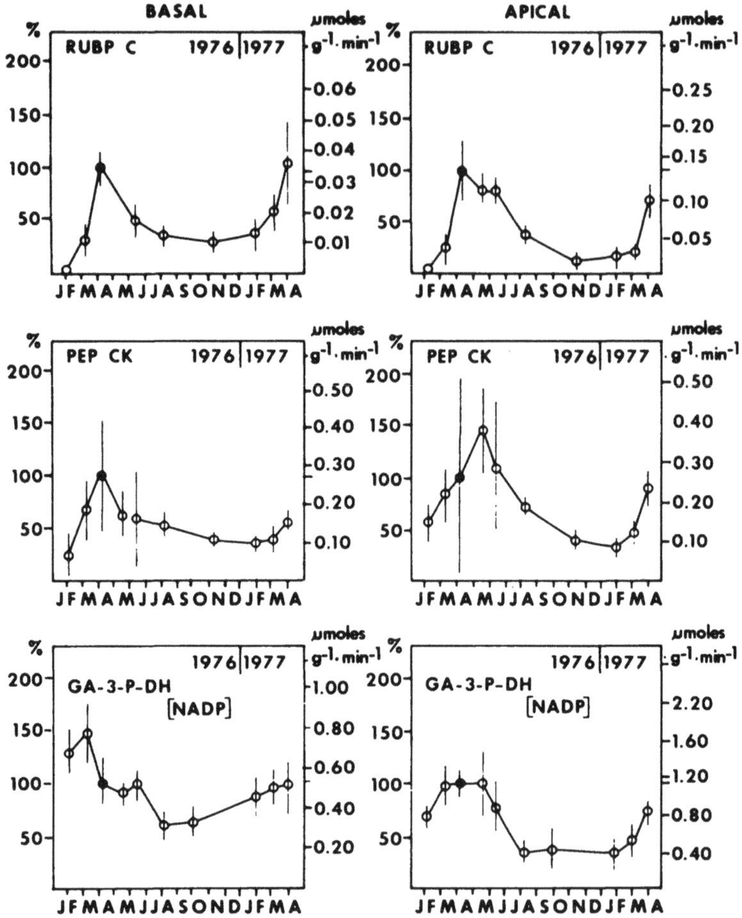

Fig. 7: Seasonal changes of enzyme activities (measured at 25 C) in basal (young) and apical (old) parts of fronds of Laminaria hyperborea. April values are set as 100%, bars indicate standard deviation. RuBP-C: ribulose-1,5-bisphosphate- carboxylase; PEP-CK: phosphoenolpyruvate-carboxykinase; GA-3-P-DH(NADP): NADP-dependent glyceraldehyde-3-phosphate- dehydrogenase. Abscissa: abbreviations mark the months, beginning with J = January (modified after 19).

Therefore, translocation of assimilates in Laminaria has both a quantitative and qualitative significance for the formation of a new thallus. But, one has to take into account the fact that young tissues of all Laminariales tested so far exhibit high activities of

PEP-carboxykinase, not only those which are characterized by seasonal growth periods. Measurements of several enzymes involved in carboxylation and mannitol biosynthesis have been carried out for two years in order to clarify whether or not a seasonal adaptation exists at the enzyme level (18,19). As shown in Fig. 7, the enzymes involved in carboxylation and maintenance of a reduction potential exhibit peaks in March through May. But one has to consider that enzyme studies are normally made under standard conditions, for instance, at 25 C. Temperature in vivo, however, changes drastically in the course of the year. Consequently, these data were recalculated on the basis of the in vivo temperatures and the activation energies of the enzymes. This correction led, as could be expected, to a maximum activity in summer. It should be mentioned that the maximum activity for mannitol-P-dehydrogenase was not reached until September, which supports the idea that mannitol synthesis is highest at that time of the year (19). Figure 7 also shows that for both young and old parts of the frond a compensatory increase in the amount of enzymes takes place in early spring when enzymatic activity should be lowest because of the low temperature in the sea.

References

1. Akagawa, H., Ikawa, T., Nisizawa, K.: Bot. Mar. 15, 126-133 (1972).

2. Akagawa, H., Ikawa, T., Nisizawa, K.: Plant Cell Physiol. 13, 999-1016 (1972).

3. Bean, R.C., Hassid, W.Z.: J. Biol. Chem. 212, 411-425 (1955).

4. Bidwell, R.G.S., Levin, W.B., Shephard, D.C.: Plant Physiol. 45, 70-75 (1970).

5. Bird, I.F., Cornelius, M.J., Keys, A.J., Whittingham, C.P.: Phytochemistry 13, 59-64 (1974).

6. Bisalputra, T.: In Algal Physiology and Biochemistry (W.D.P. Stewart, ed.), Blackwell, Oxford, pp. 124-160 1974

7. Chapman, A.R.O., Markham, J.W., Lüning, K.: J. Phycol. 14, 195-198 (1978).

8. Heber, U.: Ann. Rev. Plant Physiol. 25, 393-421 (1974).

9. Hellebust, J.A., Haug, A.: Can. J. Bot. 50, 177-184 (1972).

10. Howard, R.J., Gayler, K.R., Grant, B.R.: J. Phycol. 11, 463-471 (1975).

11. Joliffe, E.A., Tregunna, E.B.: Phycologia 9, 292-303 (1970).

12. Kremer, B.P.: Untersuchungen zur Biochemie und Physiologie von Mannit in benthischen marinen Algen. Dissertationsschrift, Universität Bonn 1973

13. Kremer, B.P.: Can. J. Bot. 56, 1655-1659 (1978).

14. Kremer, B.P.: In, Handbook of Phycological Methods. (J.A. Hellebust, J.S. Craigie, eds.) Cambridge Univ. Press, Cambridge, pp. 269-283 1978

15. Kremer, B.P., Schmitz, K.: Z. Pflanzenphysiol. 68, 357-363 (1973).

16. Kremer, B.P., Vogl, R.: Phytochemistry 14, 1309-1313 (1975).

17. Kremer, B.P., Willenbrink, J.: Planta 103, 55-64 (1972).

18. Küppers, U.: Enzymologie der CO_2-Fixierung bei Laminaria hyperborea: Jahresperiodische, umweltbezogene Veränderungen von Enzymaktivitäten. Dissertationsschrift Universität Köln 1978

19. Küppers, U., Weidner, M.: Planta 148, 222-230 (1980).

20. Lüning, K.: Mar. Biol. 2, 218-223 (1969).

21. Lüning, K.: European Mar. Biol. Symp. 4, 347-361 (1971).

22. Lüning, K., Schmitz, K., Willenbrink, J.: Mar. Biol. 23, 275-281 (1973).

23. Nordhorn, G., Weidner, M., Willenbrink, J.: Z. Pflanzenphysiol. 80, 153-165 (1976).

24. Percival, E., Smestad, B.: Phytochemistry 11, 1967-1972 (1972).

25. Raven, J.A.: Biol. Rev. 45, 167-221 (1970).

26. Schmitz, K.: Zum Assimilattransport in marinen Algen: Struktur und Funktion der Siebelemente bei den Laminariales. Habilitationsschrift Universität Köln 1976

27. Schmitz, K., Lüning, K., Willenbrink, J.: Z. Pflanzenphysiol. 67, 418-429 (1972).

28. Schmitz, K., Srivastava, L.M.: Can. J. Bot. 53, 861-876 (1975).

29. Weidner, M., Küppers, U.: Planta 114, 365-372 (1973).

30. Willenbrink, J., Kremer, B.P.: Planta 113, 173-178 (1973).

31. Willenbrink, J., Kremer, B.P., Schmitz, K., Srivastava, L.M.: Can. J. Bot. 57, 890-897 (1979).

32. Willenbrink, J., Kremer, B.P., Schmitz, K., Weidner, M.: Ber. deutsch. Bot. Ges. 92, 157-167 (1979).

33. Willenbrink, J., Rangoni-Kübbeler, Tersky, B.: Planta 125, 161-170 (1975).

34. Winkenbach, F., Parthasarathy, M.V., Bidwell, R.G.S.: Can. J. Bot. 50, 1367-1375 (1972).

35. Yamaguchi, T., Ikawa, T., Nisizawa, K.: Plant Cell Physiol. 7, 217-229 (1966).

SEASONAL PATTERNS OF CARBON ASSIMILATION AND UTILIZATION BY KELPS[1]

K.H. Mann, A.R.O. Chapman and J. Gagné
Department of Biology, Dalhousie University
Halifax, N.S., Canada B3H 4J1

Earlier work had shown that in St. Margaret's Bay, Nova Scotia, where nutrients limit growth in summer, Laminaria longicruris shows two major adaptations to maximizing annual production. During winter, when nutrients are abundant, it stores nitrate to a concentration 28,000 times ambient, and uses the stored nitrate to prolong growth into late spring and early summer. In late summer it stores carbon in the form of mannitol and laminaran, then uses it for growth in fall, when nutrients are available but light is limiting.

It is now shown that this pattern is not invariate. Studies at a site where nutrients were never limiting showed that under these conditions L. longicruris had negligible storage of either nitrate or carbohydrate. Instead variation in growth rate paralleled variation in light flux. At a site where nutrients were limiting for a short period in summer, there was winter storage of nitrate but greatly reduced storage of carbohydrate in summer.

A study of Laminaria solidungula in the Arctic showed that its growth was strongly controlled by availability of nitrogen and was out of phase with variation in light flux. On the other hand, Laminaria pallida in South Africa experiences no nutrient limitation so that variation in growth rate follows the seasonal light pattern. It is concluded that for all these species photosynthesis tends to be maximal in summer and minimal in winter, and that growth follows the same pattern provided nutrients are

[1] The full text of Dr. Mann's paper delivered at Bamfield Meetings appears in: Primary Productivity in the Sea. (P.G. Falkowski, ed.). Brookhaven Symposia in Biology 31 Plenum, N.Y., pp. 363-380 1980 (ed. note).

Synthetic and Degradative Processes in Marine Macrophytes

not limiting. When they are limiting, nutrient storage and carbon storage are used as devices for maximizing growth. It is not known to what extent the presence or absence of storage phases in L. longicruris is genetically controlled.

DISCUSSION

LÜNING: Just a comment, Dr. Mann. Nitrogen is certainly a vital factor for growth, but I wonder whether nitrate is directly regulating and inducing these seasonal patterns of the kelp plants? There are indeed different strategies, but one can also say that there is a general pattern of seasonal periodicity in kelp plants. Some of these patterns are: that they form the blade, or the majority of the blade at a very early time in the year. In the Arctic Laminaria solidungula, the blade is formed already in darkness so that, during the short period of light that it has, the blade is ready for capturing light quanta. In Laminaria longicruris, where there is not so much nutrient limitation, there is a seasonal pattern of a minimum growth of 0.5 cm per day in November and a maximum in April, May and June. So there is a seasonal pattern even there. Another thing which you did not mention, most of these plants make sori in Autumn, when the growth rate goes down. I think you are right, nitrate is determining the seasonal patterns, but maybe only as one of vital factors -- like Liebig's conditions, a minimum factor. Tony Chapman was able to increase the growth rate in a kelp bed, where there was nitrate limitation, by pouring nitrate into the water. That was a good experiment. But what we need are tank experiments where one can show whether nitrate is inducing these things or not. Somehow I doubt that these complicated plants have survived for millions of years just by reacting to nitrate concentrations in the water.

SRIVASTAVA: In your growth measurements, did you also measure the weight of the plants or was it only the increase in surface area, cm^2/day?

MANN: The first year of sampling -- the data that I have shown -- did not take into account the biomass production. But there is a second year of data, which is nearing its end of collection, which includes biomass production. It looks to me very much as if the plants that cope with nitrogen limitation do so so effectively that they produce almost as much as those that have no nitrogen limitations. Certainly their summed growth increments are almost as great in St. Margaret's Bay as they are at the site at the south end of the province. This is of interest because, there is a reference made to Tony Chapman's experiment in which he increased growth rate in summer by adding nutrients in St. Margaret's Bay. There is a snag to this and that is that, while those plants were growing in summer, they were not building up carbon reserves and so they did not grow in fall when their colleagues -- the ones growing on the carbon reserves -- did. It looks as if the combined strategies of storing nitrogen and storing carbon can lead to as much production in the course of a year as by plants which are not subject to nutrient limitation in summer.

BIDWELL: There is a parallel here from our work with Chondrus in culture. If you add a substantial amount of nitrogen you do indeed greatly increase the growth of the plants but, as you would guess, you do not get very large carbohydrate formation at the same time, and since we are interested in carbohydrate this is really counter-productive. However, you can take the plants, make them grow fast by adding nitrogen, and by a subsequent treatment in which you put the plants under rather stressful conditions for a short period of time, you can get them to convert a large proportion of the carbon that they laid down, much of it as protein, back into carbohydrates. In other words, it seems that they do have the capacity to backtrack at that point and prepare for an expected bad situation. Well, then you simply harvest them when they have a maximum carbohydrate level. You can indeed benefit from both sides of that question, first, by adding the nitrogen to make them grow big, then withholding it to make them convert proteins into carbohydrates.

I was also interested in your observation about leakage. We did some experiments with a unicellular freshwater alga, Oocystis, growing in a rather eutrophic situation in Hamilton Harbour. We were interested in its photosynthesis and so called photorespiration. We carefully purified the cultures, made them essentially uni-algal and axenic, as nearly as we could, and found that there was no leakage of carbon and no detectable photorespiration. This was of concern because Harris had done many experiments on algae from Hamilton Harbour and had found that there was a lot of photorespiration and a lot of leakage. So we thought if we added Hamilton Harbour back to these algae they might be happier. And sure enough, as soon as we added back some of the bacteria, and other much smaller algae or algal-like organisms that were in there, the Oocystis then began not only to leak carbon in large amounts but this carbon was then oxidized by these symbiotic organisms, and we got the appearance of photorespiration. It was not true photorespiration, in fact, but it would seem that, in the absence of associated heterotrophic organisms, the Oocystis did not leak any carbon, it just kept it inside. It suggests that there may be inter-organism mechanisms which could control leakage. This may be one of the reasons why so many widely different results have been obtained, sometimes from the same organism, regarding the leakage of glycolate and other similar organic compounds into the sea.

MANN: Well, that is an interesting angle. From the papers I have seen, no one seems to have stated under what nitrogen conditions they carried out experiments to determine carbon leaks; it does seem to me that there is a lot of circumstantial evidence pointing to release of carbon when growth is limited by nitrogen shortage and not at other times. If the experiments are done with an abundant nitrogen then there may be no release.

BIDWELL: The same situation applies to lichens where the export of carbon from algae increases dramatically at the time the lichens are undergoing their most active growth or sporulation and again there would seem to be some inter-organism communication system which suggests the control of leakage -- well, leakage may be the wrong word -- loss of excess carbon perhaps.

LOBBAN: Dr. Mann, how does reproduction tie into your story? When do these different plants reproduce and how much carbon goes into it? I wonder if that is where your 30% carbon went.

MANN: Tony Chapman claims to have calculated that no more than about 1% of the carbon budget of these algae goes into reproduction. That is the reason we left it out.

LOBBAN: I wonder if they keep growing at the same rate, while they reproduce?

MANN: Klaus Lüning made the point that fall is the time when they produce their sori and fall is the time they all show a dip in growth. But we have taken our cue from Tony Chapman whose estimates of the biomass of products released do not allow for much more than 1%. These actively growing plants replace their blade 5 times or more per year. Thus, the biomass of a blade is no more than 20% of total production, and since the biomass of the spores released is only 5% of the biomass of the blade then you are down to 1% of the annual production.

LOBBAN: They also tend to lose quite a lot of tissue along with spores and the whole sorus goes as well. Also, some of them reproduce all through the winter as well, do they not?

MANN: I have very little doubt that the dip in growth in fall is associated with the reproduction phase, but what is so striking to me is that some cease growth completely in the fall and some grow like seeds -- drawing down their carbon reserves.

SCHMITZ: Could you tell us where in the cell or the plant mannitol is being stored?

MANN: No, sorry.

DUNCAN: In the measurements of nitrogen in the ambient water was it always just nitrate, or were there any measurements made of organic nitrogen as well?

MANN: No. It is all inorganic nitrogen, but it included ammonia, nitrate and nitrite. Nobody has looked at organic nitrogen in the water.

DUNCAN: I ask this because, there was an 11 year study from Plymouth in the English Channel showing that when the inorganic nitrogen was down the organic nitrogen was up and, if it was averaged out over a long period of time, total nitrogen was essentially the same all the time. They were looking primarily at planktonic organisms and found that those that could utilize organic nitrogen did so and increased their population while the other plankton, which used the inorganic nitrogen, were going down. There is a possibility of the same thing happening with the larger algae.

MANN: I have heard this to explain dinoflagellates replacing diatoms. Dinoflagellates seem to be able to use organic nitrogen. For macroalgae, Tony Chapman wanted to test the idea that the only way these algae could survive in the Arctic was by taking up organic materials, that is, by being heterotrophic, and he felt that he had demonstrated adequately that there is enough photosynthesis to account for the observed production. All you need to invoke for winter growth is the utilization of carbon reserves.

DUNCAN: And no uptake! Maybe both!

MANN: There are people in oceanography at Dalhousie, who work with organic nitrogen in the water who say that most of it is refractory, in the sense that bacteria can not get at it. So they wonder what else does?

DUNCAN: Naval Antia showed that a number of planktonic organisms other than dinoflagellates take up and use amino acids.

MANN: Yes, we had a colleague from Southhampton with us for a term, and he maintained that as soon as any amino acid was released into the water there were bacteria ready to take it. And that they were much more efficient at taking it up than any plant that you could name.

DUNCAN: Then they can release ammonium which is available to the plants.

MANN: Yes, sure. But then you are back to inorganic cycles.

ENGELMANN'S THEORY: THE COMPELLING LOGIC

J. Ramus

Botany Department and Marine Laboratory, Duke University
Beaufort, North Carolina, USA 28516

The coincidence of wavelengths of maximum transmission through the marine water column (green light) and maximum photosynthesis by red seaweeds (green light) crystallized a compelling logic, the theory of complementary chromatic adaptation. The theory was originally proposed by T.W. Engelmann (1,2) after he produced crude photosynthetic action spectra for numerous algae by an ingenious O_2 bioassy and a projected microspectrum. The action spectra of the green algae and diatoms were relatively inactive at wavelengths between the two major absorption peaks of chlorophyll, the spectral region here designated the "green window." However, the action spectra of red algae were remarkably different: photosynthesis was sensitized in the green window, the spectral region of phycoerythrin absorption. The principle presumably limiting the lower extreme of seaweed vertical distribution was the ability to capture light in the green window. Thus the color of red seaweeds carrying large quantities of phycoerythrin relative to other antenna pigments complements the color of the submarine light field. Engelmann's theory was *sensu strictu* genotypic, making no allowance for phenotypic color plasticity, and applied solely to seaweeds. In addition, the theory was *a priori*: the red seaweeds had evolved for effective light absorption in the deep sea.

Engelmann's theory here shall be distinguished from its extension, formulated by his student Gaidukov (3), to include phenotypic adjustment in the pigment systems of prokaryotic algae to colored light. He observed that blue-green algae became green in red light, blue in green light, yellow in blue-green light and blue in yellow light. Oddly enough, Engelmann's first theory was formulated for seaweeds, yet his observations were based on only two filamentous red (*Callithamnion* and *Ceramium*) and possibly one green (*Cladophora*) seaweed; the bulk of the species observed were diatoms and freshwater green algae. Nevertheless, subsequent work by others fleshed out the theory, although it was subject to criticism. The controversy is exhaustively reviewed in the treatise on photosynthesis by Rabinowitch (4,5) and will not be recapitulated here. A full exposition of the theory provides that the green seaweeds (carrying only chlorophylls) occupy the most restricted vertical distribution (a narrow zone near the surface), the browns (carrying fucoxanthin) an intermediate distribution and the reds (carrying phycoerythrin) the greatest vertical distribution. The implications were clear: the vertical distribution of seaweeds is imposed solely by the color of the submarine light field and there exists no other mechanism for filling the green window save insertion of biliproteins. The logic is so manifest that the theory has persisted in the face of considerable contrary evidence. For example, believing Engelmann fundamentally correct, the esteemed Rabinowitch (5) summarized: "Would it not be strange if the appearance of orange or red pigments in deep-water algae would only be a coincidence, and these pigments were helpless in performing the task so obviously set to the plants by the character of the light in which they live -- to catch and utilize for their maintainance and propagation radiations in the middle of the visible spectrum, which are the only ones to reach them in some intensity."

To detract from Engelmann is a mistaken enterprise. His experiments in photobiology were bold for the time and elegant in design. Indeed, the experiments have withstood the scrutiny of repetition in the subsequent century. Yet Engelmann's theory as it flowed from the primary data was just that, a theory, an axiom he never tested. The inference was so strong that in the last century the theory has been elevated to principle without appropriate verification. Now the burden of testing Engelmann's hypothesis, or of devising alternative hypotheses, is left to us.

In the broadest sense, we are seeking mechanisms which structure seaweed communities. Seaweeds are a life form, and their assignment to three divisions of the plant kingdom acknowledges their separate phylogenies. It follows that unique genetic repertoires have in the course of evolution converged in niches broadly similar in terms of physical properties and biological interactions. The distinct genetic repertoires are manifest biochemically in the structure of the photosynthetic unit (PSU), primarily in the kinds of pigments in the antenna since we know the rest of the PSU to be evolutionarily conservative (6). Engelmann of course knew nothing of the modern concept of the PSU, but he knew of the major differences in antenna pigments giving red, green and brown seaweeds. The theory of complementary chromatic adaptation then serves as the basis for the formulation of a specific (and testable) question: does the spectral distribution of the submarine light field structure seaweed communities? It is this question which we shall now systematically address. However, the question is immediately exacerbated by the simple fact that the absorption of light by a water column is wavelength dependent, viz. change in light quantity with depth is always accompanied by change in spectral distribution. Further, water columns vary in the wavelength of maximum transmission, hence the optical codification of coastal and oceanic waters by Jerlov (7). Any attempt to assess the structuring role of spectral distribution must separate color from quantity and examine their effects independently.

There are at least two broad approaches to the analysis of community structure. In the first, we assume that Nature has already tested the hypothesis - we need merely to analyse existing communities for verification or negation. The analysis may be either functional or descriptive (8). In the functional analysis, component species are experimentally manipulated, usually removed, then the impact on total community structure is appropriately assessed over time. The works of Paine and Vadas (9), Dayton (8), Lubchenco (10) and Sousa (11) leave us with a sobering thought - seaweed community structure is a consequence of many mechanisms acting in concert, viz. physical factors, plant-plant interactions and plant-animal interactions. The notion in question here, that light, and specifically the color of light, is the sole mechanism structuring seaweed communities is summarily naive. There is no question that light plays a role. This is easily demonstrated by removing canopy species, upon which the obligate understory species die out and the newly opened free space is occupied by fugitive species (8). Whether it is the quantity of light (photon flux density, PFD) or quality (spectral distribution) remains moot.

Descriptive analysis of community structure is here the bathymetric quantification of relative species abundances, or perhaps more appropriate, class abundances, insofar as we address class specific PSU structure. To our purpose, we are immediately confronted by a profound dilemma - in what units are abundances to be measured: the number of individuals per species, % cover per species, biomass per species, productivity per species? Said otherwise the question is: on which of these parameters does natural selection operate? Although crucial to the discussion, this is not to be resolved here.

Nevertheless, there are several bathymetric analyses of seaweed vertical distribution which are of use to this treatise. Larkum et al. (12), investigating a cliff in clear oceanic water of the Mediterranean, found that brown seaweeds dominated the upper 15 m, greens dominated from 15 to 75 m (the bottom) and the reds made little contribution to biomass at any depth. To extrapolate the occurence of seaweeds below 75 m, the zonation in a submarine cave was quantified. Here, siphonaceous green seaweeds penetrated some distance inside on the near-vertical walls and were eventually replaced by encrusting red species. The transition from green to red species occurred at the PFD equivalent to a depth at 80 m outside the cave, with reds becoming sparse at the equivalent of 120 m. In this instance, red seaweeds penetrated "deeper" than other types. However, Larkum et al. (12) calculated that the proportion of light available for photosynthesis by red and green seaweeds at various depths in these oceanic Jerlov type II (7) waters is virtually the same down to 150 m, well below the compensation depth for all species. They also noted that in the submarine cave and on the north-facing cliff, light is already confined to the blue wavelengths well before the transition from green to red seaweeds occurs. Thus, Larkum et al. (12) concluded that light quantity rather than spectral distribution was the major regulating factor for the pattern of vertical distribution.

Doty et al. (13) investigated vertical distribution of Hawaiian seaweeds seaward of the algal ridge. The ratio of brown to green species below both 10 m and 50 m was 0.5 and below 90 m was 0.0. The ratio of red to green species below 10 m was 3.0 , below 50 m was 2.2 and below 90 m was 0.7.

Thus, there was a progressive enrichment for green species with depth. Titlyanov (14), working in the turbid waters of the Gulf of St. Peter the Great, reported the ratio of brown to green species at depths less than 2 m to be 2.1 and at depths exceeding 12 m to be 1.7, while the ratio of red to green at depths less than 2 m to be 3.4 and greater than 12 m to be 3.7. He concluded by stating that the species composition at the greatest depths of habitation is not correlated with taxonomic groups of specific pigmentation i.e., classes. Schneider (15), studying the spatial distributions of seaweeds on the continental shelf of the Carolinas, reported that 25% of the green species, 16% of the brown and 28% of the reds occurred at depths greater than 50 m.

Early on, Oltmanns (16) proposed that the vertical distribution of seaweeds was controlled by light quantity alone. Titlyanov (14) investigated a contiguous shallow grotto and submarine cliff. The situation appeared an ideal test: the grotto represented a gradient in light quantity alone while the adjoining submarine cliff represented both a gradient in light quantity and a change in spectral distribution. The species common to neighboring well-lit areas are absent inside the grotto. The grotto supports shade-tolerant species which grow nearby in deep crevasses or underwater cliffs. A critical observation is that the red seaweeds <u>Ptilota filicina</u> and <u>Symphocladia latiuscula</u> living in the grotto and in cliff crevasses near the surface appear again in the open at a depth of 10-15 m. Thus, Titlyanov (14) concludes that their occurrence is a function of light quantity alone and the change in spectral distribution is a minor contributor to the vertical stratification of seaweeds. It should be noted here that the spectral distribution of light along a horizontal line of sight is skewed in favor of shorter wavelengths, viz. sidewelling light entering a submarine grotto is predominantly blue (17)!

Always irksome in this context are the pavement reds, which, when occurring in a flora, appear to persist to depths where no other seaweeds exist. Goreau (18) has speculated that the success of encrusting reds at low PFD is due to a slow rate of attrition rather than to exceptional utilization of available light. Their calcareous nature and prostrate habit is a deterrent to both grazing and physical erosion. In conclusion, it would appear that analyses of community structure, whether using the functional or descriptive approaches, fail to verify Engelmann's theory. However, it is not likely (for the reasons noted) that the existing analyses of seaweed community structure could either verify or negate the paradigm.

The second approach to testing Engelmann's theory is experiment designed to eliminate all structuring mechanisms save light, then isolate the effects of spectral distribution from PFD. These are tests focused specifically on the PSU, correlating its structure and function with depth. For example, Levring (19) noted that photosynthesis and PFD decrease in a logarithmic manner with depth. He reasoned that if photosynthesis at two different depths, m and n, are P_m and P_n and the corresponding PFD values are E_m and E_n, then the quotient of the slopes of the two curves, q_{m-n}, can be calculated by the relationship:

$$q_{m-n} = \frac{\log (P_m/P_n)}{\log (E_m/E_n)}$$

If photosynthesis were independent of spectral composition but dependent on PFD, then P and E would reduce proportionally with depth and q = 1. If photosynthesis were dependent on spectral composition and indpendent of PFD, then P and E would not reduce proportionally with depth and q would be greater or less than unity, $q > 1$ representing a spectral impoverishment and $q < 1$ representing a spectral enrichment. Photosynthesis (P) and irradiance (E) were measured at depths m and n where E was less than saturating. Levring compared q values in turbid coastal (green) and clear oceanic waters (blue) over the Carolina continental shelf. For Ulva lactuca in neritic waters q increased with depth from 0.92 to 1.35 (progressive spectral impoverishment) and in oceanic water q decreased from 1.18 to 0.79 (spectral enrichment). For Gracilaria foliifera in neritic waters q decreased from 0.99 to 0.72 (progressive spectral enrichment) and in oceanic water q varied erratically and the data are best interpreted as spectral indifference. For Fucus vesiculosus in both neritic and oceanic waters the q values are best interpreted as spectral indifference. These data do not conform to the expectations of the theory. Dring (20), using a computer model which calculated photosynthesis from action spectra and underwater spectral distribution data, simulated photosynthesis as a function of depth and water type for many species of seaweeds. The species of red seaweeds were best adapted chromatically to photosynthesize at all depths (including 0 m) in all except the clearest oceanic waters. The results show little correlation with patterns of vertical distributions of green, brown and red seaweeds, and Dring (20) concludes that complementary chromatic adaptation may be of little importance in determining vertical zonation of seaweeds. As if anticipating this conclusion, Drew (21) calculated an increase in photosynthetic efficiency for Udotea petiolata (a green), Peysonnelia squamaria (a red) and Padina pavonia (a brown) with depth, in that order, in the clear oceanic waters off Malta.

If in fact photosynthesis and patterns of vertical distribution are dependent primarily on light quantity and independent of light quality, then green and brown seaweeds must also have effective mechanisms for the absorption of light in the green window (500-600 nm). Brown seaweeds carry both fucoxanthin, the in vivo absorption spectrum of which extends out to 580 nm, and chl c, and effectively absorb the entire PAR spectral domain. Yokohama and Kageyama (22), comparing the in vivo absorption spectra of five shallow-water and six deep-water green seaweed species, discovered an unexpected absorption peak at 540 nm in deep-water species which was not present in shallow-water species. The pigment absorbing light in this region was identified chromatographically as siphonoxanthin. Siphonoxanthin is characteristic of many siphonaceous greens (23), so the occurrence of the pigment in species of Cladophora, Codium and Valonia came as no surprise. The fact that Ulva japonica carried the pigment is unexpected. Kageyama et al. (24) showed an efficient excitation energy transfer from siphonoxanthin to chl a by fluorescence emmission studies. In U. japonica, the efficiency of transfer at 540 nm is 50% greater than for the shallow water species U. pertusa containing no siphonoxanthin. Photosynthetic efficiency for U. japonica collected from a depth of 20 m was equal in "green light" (simulated deep coastal water) to "white light," whereas U. pertusa from shallow water was 30% less efficient in "green light" than in "white light" (22). There seems little doubt that U. japonica has evolved an efficient mechanism for the capture of green light, allowing the species to exist in spectrally limited light fields.

The siphonaceous green Ostreobium growing endolithically in the coral head Favia and beneath the outer layer of zooxanthellae is subject to a unique spectral distribution. PAR wavelengths relative to far-red are transmitted through zooxanthellae in very low ratios (25). Action spectra for Ostreobium reveal that the "red drop" does not occur near 700 nm as expected, but rather beyond 740 nm. It would appear that Ostreobium possesses chlorophyll, both antenna and trap, with absorption maxima in the far-red.

It has been demonstrated that the green window in seaweeds can be filled by chlorophyll absorption alone. The argument is a complex one and has been put forth in detail elsewhere (26). In summary, as the concentration of chlorophyll increases, the spectral region of low specific absorptance (the green window) increases more rapidly than the regions of high specific absorptance (the blue and red absorbing regions). Thus, Ulva lactuca with high concentrations of chlorophyll absorbs light almost uniformly across PAR wavelengths (26). Most seaweeds, however, are very heterogeneous systems with respect to the absorption of light. The absorptance of the siphonaceous green seaweed Codium fragile is always near 1, independent of chlorophyll concentration within natural ranges, due to the effects of heterogeneous absorption and multiple scatter (26). The utricles forming the cortical tissue enhance the capture of incident light due, among other things, to the waveguide function of the thin peripheral layer of cytoplasm in which the chloroplasts are arrayed. Thus the length of the optical path is made larger and increases the probability of photon - chloroplast encounter, independent of wavelength. Many siphonaceous greens have a cortex of similar structure, perhaps contributing to their presence in even the weakest of submarine light fields.

It is well known that antenna pigment concentration varies inversely with PFD, e.g. the pigment concentration of a species in a strong light field compared to a weak light field is lower, as is usually the ratio of accessory to primary antenna pigments. These quantitative adjustments are, within genotype, a phenotypic expression of PSU plasticity.

The PSU of red seaweeds has undergone the most intensive investigation, prompted by the phenotypic corollary of Engelmann's axiom. In this context, his theory is extended to include phenotypic adjustments to light quality and/or quantity. Aside from chl a, the red seaweed antenna contains phycoerythrin (PE), phycocyanin (PC) and allophycocyanin (APC). Here, the specifics for the model of the PSU are controversial. Initially it was thought that the role of the APC was solely to transfer excitation energy from phycobilisomes to the traps, but Larkum and Weyrauch (27) have shown that APC harvests a substantial amount of light. Further, Larkum and Weyrauch (27) maintain that the phycobiliproteins are in a sense independent of both photosystems and transfer absorbed light energy preferentially to PS II with excess energy spilling over to PS I. On the other hand, Ley and Butler (28) suggest that the phycobiliproteins are associated entirely with PS II.

Waaland et al. (29) demonstrated that adjustment of the PSU in Griffithsia pacifica is effected by PFD and not by spectral distribution. PE/chl a ratios (mg/mg) varied from 0.67 at 1000 f.c. to 1.8 at 8 f.c., the greatest variation occurring below 300 f.c. The correlation between phycobiliprotein concentration and phycobilisome density was tight, indicating that perhaps the PSU number varied here. Ramus et al. (30) reported for Chondrus crispus and Porphyra umbilicalis in situ an increase in total antenna pigment concentration and a PE/ chl a and PC/ chl a enrichment with decreasing PFD. Plants suspended at varying depths in the water column (where both spectral distribution and PFD change) were compared with plants from intertidal habitats where only the PFD changed. Antenna pigment concentration varied in a similar manner, leading to the conclusion that these red seaweeds were responding to PFD alone. The adjustment of the PSU was reversible and did not require cell division. Rhee and Briggs (31) reported that the chlorophyll variation in Chondrus crispus was not correlated with depth, but PE did increase with depth. The evidence indicated that the phycobiliprotein/ chl a variations are correlated with PFD, yet it is not known whether different spectral distributions of equal quantum flux density will produce the same adjustments.

For green seaweeds, an increase in total chlorophyll and chl _b_/ chl _a_ ratio is correlated with a decrease in PFD (22,26,30,32). Chl _b_/ chl _a_ ratios vary from 0.44 in high PFD environments to 1.12 in low PFD environments, while total chlorophyll can vary up to 15-fold in _Ulva_ and 8-fold in _Codium_ (26). Apel _et al._ (33) released PS II from the PSU of _Acetabularia mediterranea_ by detergent treatment and showed it to be enriched in chl _b_. In that the chlorophyll-protein complex photoreduced dichlorophenolindophenol, but not $NADP^+$, they concluded it was PS II. Siphonoxanthin transfers absorbed energy to chl _a_ in _Ulva japonica_ (24), yet its photosystem association is unkown. Molar ratios of siphonoxanthin/ chl _a_ range from 0.25 in _Codium fragile_ to 0.35 in _Valonia macrophysa_. Carotenes, of course, are present in abundance in the chloroplasts of green seaweeds (23), but as evidenced from enhancement spectra (34,35) and fluorescence excitation measurements (24), they may participate less efficiently in energy transfer.

The three major antenna pigments in the brown seaweeds _Ascophyllum nodosum_ and _Fucus vesiculosus_ increase with decreasing PFD. However, chl _c_/ chl _a_ ratios remained constant, while fucoxanthin/ chl _a_ actually declined; an increase in fucoxanthin was not as great as in chl _a_ for these seaweeds (36).

Mishkind and Mauzerall (6) have specifically addressed the notion of light-mediated structural modification of the PSU. Their definition of PSU size is a functional one and consistent with that of Emerson and Arnold (37): the molar ratio of chlorophyll to O_2 formed in a single saturating turnover flash (in this case the flashes were pulsed at 100 ms intervals). The average PSU size for the green seaweeds is similar to that of _Chlorella_ and higher plants, _viz._ 1500-2500 chl/ O_2 and about 1/3 that size for brown and red seaweeds. In the browns and reds, a large complement of non-chlorophyll accessory pigment is compensated for by a reduction in chlorophyll content, but by no means do accessory pigments replace all the chlorophyll. The data of Mishkind and Mauzerall (6) suggest that _Ulva lactuca_ adjusts its PSU to low PFD in the same manner as higher plants, i.e. by increasing the number of PSUs rather than antenna size. For _Ulva_, PSU size is independent over a 4-fold range of chlorophyll concentration. For _Porphyra umbilicalis_, PSU size increases with chlorophyll content as PFD decreases as does apparently the number of "traps."

There are in some cases structural modifications of the chloroplast or cell which accompany adjustment in the PSU. Unfortunately, the effects of structural modifications on quantum capture capacity or photosynthesis have not been quantified, leaving their significance to be demonstrated. Ulva fenestrata taken from the dim recesses of a submarine grotto had larger chloroplasts with denser thylakoid packing than counterparts growing in the intertidal (38). An increase in phycobilisome density was correlated with a decrease in PFD in Griffithsia pacifica (29), while the size of phycobilisomes remained constant. Chloroplasts of the siphonaceous green seaweed Halimeda tuna growing at 16 m were larger than those of the same species growing near the surface; however, no difference was noted in the lamellar system (39). Deep-water plants have thinner segments, larger utricles and thinner cell walls than shallow-water plants, and the chloroplasts are distributed over a wider area but in fewer layers (40).

Variations in the PSU have been correlated with in situ photosynthesis (26,36,41). Comparing photosynthesis normalized to chl a the apparent efficiency of low PFD acclimated seaweeds differs from high PFD acclimated counterparts. At the surface, the high PFD acclimated seaweeds are most efficient, while at depth the low PFD seaweeds are most efficient. When in situ photosynthesis is normalized to biomass, the compensation depth for low PFD seaweeds is always greater. Thus, an adjustment in the PSU to low PFD enhances photosynthetic performance at depth, and this is true for all seaweed classes.

Given the richness of mechanisms for the capture of light in the green window described here, it appears that green and brown seaweeds, as well as reds, have in the course of evolution undergone complementary chromatic adaptation. Where then was Engelmann's logic erroneous? Action spectra for photosynthesis are traditionally produced with monochromatic light. The spectra have been very useful for deducing the mechanisms of photosynthesis, e.g. the celebrated "red drop" and "Emerson enhancement."

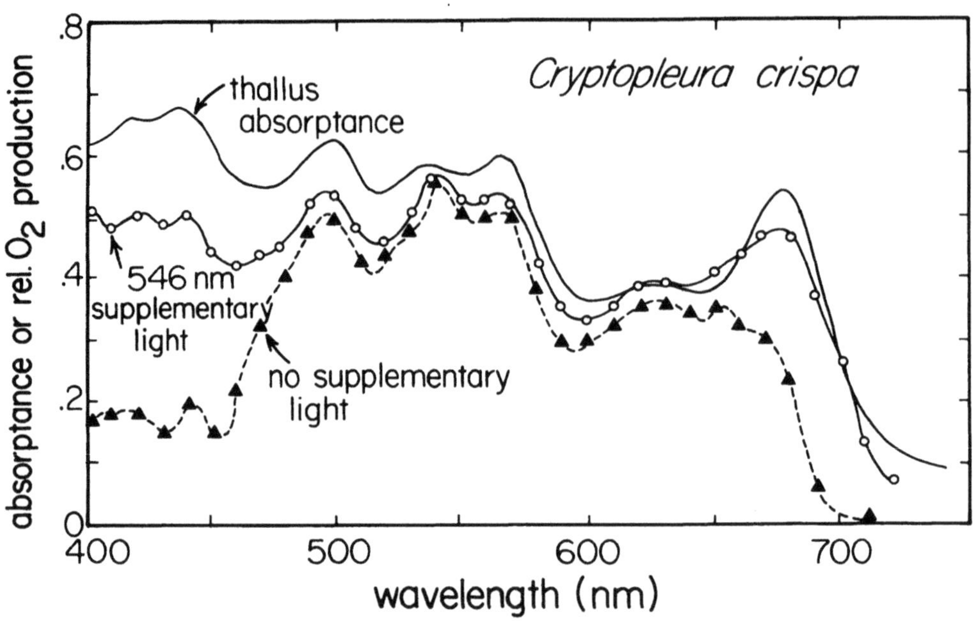

Fig. 1: Absorptance and action spectra (with and without 546 nm supplementary background light) for the red seaweed Cryptopleura crispa. Redrawn from Fork (35).

But for ecological purposes, the classical method may be misleading. Submarine light fields, at least at or above the compensation depth for seaweeds (here set at two to three orders of magnitude below surface PFD) are always broadband (42). Further, spectral distribution varies as a function of water type. In Jerlov I (clear oceanic) waters, the spectral distribution at compensation depth is 400-520 nm with maximum PFD at 480 nm (blue-green light). At the other extreme, in Jerlov 9 (turbid neritic) waters, spectral distribution at compensation is 500-680 nm with maximum PFD at 580 nm (yellow-orange light). Thus submarine light fields are not as spectrally pure as conceived in Engelmann's theory.

In that the spectral distribution of submarine light above the photosynthetic compensation depth is always broadband and of variable color, action spectra might more appropriately be produced with supplementary weak background light. When done so, the results are revealing. For example, in the red seaweed Cryptopleura crispa the action spectrum produced with 546 nm supplementary light more closely "fits" the absorptance spectrum than one produced in the absence of supplementary light (Fig. 1, reference 35). It can be inferred from action spectra with background light near 680 nm of the green Ulva and the brown Endarachne binghamiae that the decline in photosynthesis in the spectral range of the green window is not so severe as in monochromatic light (35). Had Engelmann produced his photosynthetic action spectra for red seaweeds in supplementary background light, he may never have logically inferred the theory of complementary chromatic adaptation.

Acknowledgements

I wish to thank Matt Dring and Klaus Lüning for stimulating discussion, and Ken Mann and Anne Walter for critically reading this manuscript.

References

1. Engelmann, T.W.: Bot. Ztg. 41, 18-29 (1883).
2. Engelmann, T.W.: Bot. Ztg. 42, 81-93 (1884).
3. Gaidukov, N.: Ber. deutch. Bot. Ges. 21, 484 (1903).
4. Rabinowitch, E.I.: Photosynthesis, Vol. I., Interscience, New York (1945).
5. Rabinowitch, E.I.: Photosynthesis, Vol. II., Interscience, New York (1951).
6. Mishkind, M., Mauzerall, D.: Mar. Biol. 56, 261-265 (1980).
7. Jerlov, N.G.: Rep. Swedish Deep-Sea Exped. 3, 1-59 (1951).
8. Dayton, P.K.: Ecol. Monogr. 45, 137-159 (1975).
9. Paine, R.T., Vadas, R.L.: Limnol. Oceanogr. 14, 710-719 (1969).
10. Lubchenco, J.: Am. Nat. 112, 23-39 (1978).
11. Sousa, W.P.: Ecol. Monogr. 49, 227-254 (1979).
12. Larkum, A.W.D., Drew, E.A., Crossett, R.N.: J. Ecol. 55, 361-371 (1967).
13. Doty, M., Gilbert, W.J., Abbott, I.A.: Phycologia 13, 345-357 (1974).
14. Titlyanov, E.: Mar. Biol. (U.S.S.R.) 1, 3-12 (1976).
15. Schneider, C.W.: Bull. Mar. Sci. 25, 133-151 (1976).

16. Oltmanns, F.: Morphologie und Biologie der Algen, Vol. II. Fisher, Jena (1905).

17. McFarland, W.N., Munz, F.W.: Vision Res. 15, 1063-1070 (1975).

18. Goreau, T.F.: Ann. N.Y. Acad. Sci. 1091, 127-167 (1963).

19. Levring, T.: Bot. Mar. 11, 72-80 (1968).

20. Dring, M.J.: Limnol. Oceanogr. 26, 271-284 (1981).

21. Drew, E.A.: Proc. Intl. Seaweed Symp. 6, 151-159 (1969).

22. Yokohama, Y., Kageyama, A.: Bot. Mar. 20, 433-436 (1977).

23. Kleinig, H.: J. Phycol. 5, 281-284 (1969).

24. Kageyama, A., Yokohama, Y., Shimura, S., Ikawa, T.: Plant and Cell Physiol. 18, 477-480 (1977).

25. Halldal, P.: Biol. Bull. 134, 411-424 (1968).

26. Ramus, J.: J. Phycol. 14, 352-362 (1978).

27. Larkum, A.W.D., Weyrauch, S.K.: Photochem. Photobiol. 25, 65-72 (1977).

28. Ley, A.C., Butler, W.L.: Biochim. Biophys. Acta 462, 290-294 (1977).

29. Waaland, J.R., Waaland, S.D., Bates, G.: J. Phycol. 10, 193-198 (1974).

30. Ramus, J., Beale, S.I., Mauzerall, D., Howard, K.L.: Mar. Biol. 37, 223-229 (1976).

31. Rhee, C., Briggs, W.R.: Bot. Gaz. 138, 123-128 (1977).

32. Keast, J.F., Grant, B.R.: J. Phycol. 12, 328-331 (1976).

33. Apel, K., Bogorad, L., Woodcock, C.L.F.: Biochim. Biophys. Acta 387, 568-574 (1975).

34. Haxo, F.T., Blinks, L.R.: J. Gen. Physiol. 33, 389-422 (1950).

35. Fork, D.C.: In Photosynthetic Mechanisms in Green Plants, pp. 352-361. Publ. No. 1145 NAS-NRC, Washington, D.C. 1963

36. Ramus, J., Lemons, F., Zimmerman, C.: Mar. Biol. 42, 293-303 (1977).

37. Emerson, R., Arnold, W.: J. Gen. Physiol. 15, 391-420 (1932).

38. Titlyanov, E.A., Mashansky, V.F., Glebova, N.T.: Bot. Zh. S.S.S.R. 59, 1553-1558 (1974).

39. Colombo, P.M., Orsenigo, M.: Phycologia 16, 9-17 (1977).

40. Colombo, P.M., Orsenigo, M., Solazzi, A., Tolomio,C.: Mem. Biol. Mar. Ocean. 6, 197-208 (1976).

41. Ramus, J., Beale, S.I., Mauzerall, D.: Mar. Biol. 37, 231-238 (1976).

42. Dartnall, H.J.A.: In Vision in Fishes (M.A. Ali, ed.), pp. 543-563, Plenum Press, New York 1975

DISCUSSION

BIDWELL: What strikes me in this total analysis is that, figuratively speaking, if Engelmann had stood on his head, he would have seen the whole problem from a different perspective. What I am getting at is, instead of being concerned about the addition of pigments, as the seaweed goes deeper and deeper, for capturing the light in the green window, what he really shoud have considered is, as the seaweed comes closer to the surface it has to get rid of pigments to avoid being cooked. The only pigments that can successfully be gotten rid of are the supplementary pigments, it must not get rid of chlorophyll, because without at least some chlorophyll it cannot do any photosynthesis. So perhaps, if Engelmann had looked at the problem that way, he would never have come up with the concept of complimentary pigmentation. There were some very early experiments by Wilstätter, who found that bean plants and some other species had far more chlorophyll in them than was necessary for the amount of photosynthesis that they did; they could in fact photosynthesize very effectively if bleached, with a very small amount of pigment left, provided there was lots of light.

RAMUS: These organisms always optimize photosynthesis, they never maximize it.

BIDWELL: Right.

FOSTER: How about other ways of coping with light intensity like changing respiration?

RAMUS: We have not done any work like that personally, and I have not seen any evidence for that in the literature.

FOSTER: There are some deep growing seaweeds that seemingly grow very slowly. It seems to me that would be an entirely different strategy -- do not change anything, just grow slowly.

BIDWELL: With respect to your remark about respiration, I presume it was photorespiration you were thinking of, I do not think most algae exhibit photorespiration. They may have the metabolism for photorespiration but we have never been able to measure it in any species we have analyzed, especially under conditions that are physiologically similar to those the plant would experience in the sea. Some people have measured photorespiration in algae, but usually there has been something peculiar about their measuring system, either they have had the plant in a little container or sealed in a small amount of seawater and generally illuminated with a very bright light and so on -- the plant was under very unnatural circumstances. Most of these plants do not photorespire, so there is no way they could compensate for reduced light by photorespiration.

LÜNING: We have done some work that shows the carotenoid content increases with light intensity.

RAMUS: We have some data that says that it does, and there is some other data that says it does as well. I do not think there is any question about that.

WILLENBRINK: But I think it depends on the location of these carotenoids, not all are located in the photosynthetic unit (PSU).

RAMUS: Right.

WILLENBRINK: Most of them are outside and are well known to have a protective function, at least well known from higher plants. I doubt that this would be different in algae.

RAMUS: I think that is to protect them from superoxides.

WILLENBRINK: Also, I would like to add a bit about the colour of the red algae. It depends first on the genetical bases and secondly, on the pigment content. There are some Chondrus in Helgoland that are dark red even in the uppermost intertidal.

RAMUS: We are now finding good correlations between the composition of the photosynthetic unit and the nitrogen "loading rate" in the water column. It would seem that the composition of the photosynthetic unit is as finely tuned to ambient nitrogen as it is to light and maybe even more finely tuned to ambient nitrogen than to light. This is not unexpected because the bulk of the combined nitrogen in a seaweed cell is in fact either in a photosynthetic unit or in RuBP carboxylase. That is the bulk of the protein in the cell.

BIDWELL: Yes, we noticed this in Chondrus. The rate of maximum, possible attainable photosynthetic capacity under optimal conditions, which is never reached in the sea, is highest in March, it rises rapidly through February to late March. Then it drops precipitously throughout the summer. But the actual photosynthetic rate in the sea increases continuously until July; during that period there is a tremendous build up of photosynthetic pigments and photosynthetic proteins and presumably rapid assimilation of nitrogen.

I would also like to respond to Dr. Willenbrink's comment on colour of Chondrus. I do not know what it is like in Helgoland, but in Nova Scotia, you generally see, particularly later in the season, it is green on the surface, and it is a beautiful dark red deep down. And the same way when we grow it in culture, if we give it too much light, it just turns green. It does not die, it just turns green.

DUNCAN: In the work that you have done, particulary with the red algae, have you seen any changes in membrane structure that may be related to quantities of accessory vs. other pigments.

RAMUS: No, we have not looked at the structure of the chloroplasts.

DUNCAN: There might be some relationship here because the structure of the chloroplasts in these different groups is quite different and so are the accessory pigments.

WAALAND: We did some experiments with the red alga, Griffithsia, and found that the variation in intensity changed the number of phycobilisomes per unit area of thylakoid membranes. But we did not do any experiments with different colours of light.

RAMUS: The other person that I know who has looked at red algal chloroplast structure as a function of light intensity is Elizabeth Gantt. Several Italians have looked at the siphonaceous green, Halimeda tuna, and again that was a function of depth, where both quantity and quality were changing. But they did see changes in structure of the chloroplasts. So did Titlyanov in Vladivostok in <u>Ulva fenestrata</u>.

SEASONALITY IN LARGER BROWN ALGAE AND ITS POSSIBLE REGULATION BY THE ENVIRONMENT

K. Lüning
Biologische Anstalt Helgoland, D-2192 Helgoland,
Federal Republic of Germany

Introduction

Seasonality of kelp performance in the field is manifested in various ways, such as the seasonal course of vegetative growth and reproduction of sporophytes, or the development and fertilization of gametophytes and the subsequent appearance of young sporophytes, which in many species and at many locations occur most commonly in winter and spring. As will be shown in the first part of the present paper, photoperiod can be ruled out as a triggering factor in the development of the gametophyte in the various species of Laminariales investigated so far. Gametophyte fertility depends on certain conditions of light quality, irradiance and temperature which, in combination with the reproductive pattern of the sporophyte, lead to the production of young sporophytes mainly in winter and spring. As to the growth and reproductive activity of the sporophytes, the factors inducing and synchronizing these processes with the seasons are still largely unknown due to lack of experimental data. Some recent work discussed in the second part of this paper shows that the nutrient and temperature cycles influence the growth performance of the sporophytes.

Gametophyte Development and Its Regulation by Environmental Factors

The fact that at elevated temperatures (15-20 C) the gametophytes of the Laminariales grow vegetatively and form filamentous thalli which ultimately become macroscopic was discovered by Schreiber (46). Schreiber (46), and subsequently others (e.g., Yabu, 50), also reported that once the gametophytes are in the filamentous stage, fertility can be induced

Synthetic and Degradative Processes in Marine Macrophytes

simply by transfer of the cultures to lower temperatures, e.g., 5 C.

More recent work has shown that, in addition to lower temperature, the gametophyte culture must also be supplied with a certain quantum dose of blue light to induce fertility (34,35). As can be seen from an action spectrum, taken at 15 C, parts of the UV and the blue range, with wavelengths up to 512 nm, are effective in inducing egg production in female gametophytes, whereas the green and red range are completely ineffective (Fig. 1). In contrast to an earlier report (34), the green or red range is ineffective not only at 15 C, but also at 5 and 10 C (31).

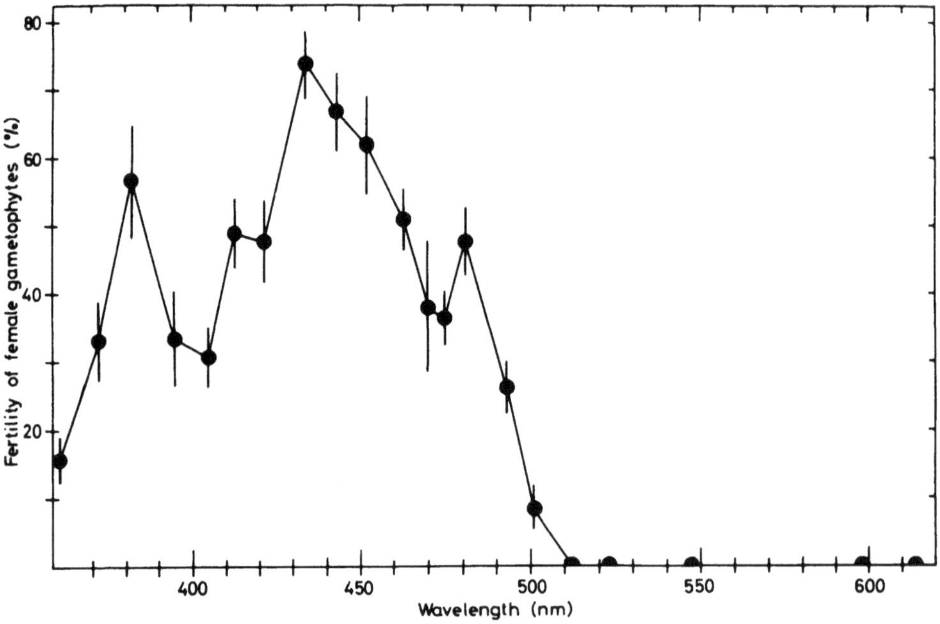

Fig. 1: Laminaria saccharina. Action spectrum for induction of egg production in female gametophytes at 15 C. Plants were grown in red fluorescent light at 15 µE m^{-2} s^{-1} for 14 days and then irradiated with 15 µE m^{-2} s^{-1} at each wavelength for 48 h. Percentage of 500 female gametophytes with eggs was determined 8 days later. UV irradiation was for 72 h at 5 µE m^{-2} s^{-1}, and results were adjusted by comparison with a control at 434 nm. Values plotted are based on results of 10 experiments. Vertical bars: 95% confidence limits. (Data from ref. 35)

The blue light requirement for the induction of fertility increases in an exponential way with increasing temperature (31). As can be seen from figure 2, which shows the blue light requirement for induction of fertility in 50% of the female gametophytes of three Laminaria spp., a quantum dose of 90 μE cm^{-2} is required at 5 C, 110 μE cm^{-2} at 10 C, and 230 μE cm^{-2} for L. hyperborea and L. saccharina, and 560 μE cm^{-2} for L. digitata at 15 C. Even higher quantum doses are required at 18 C. The

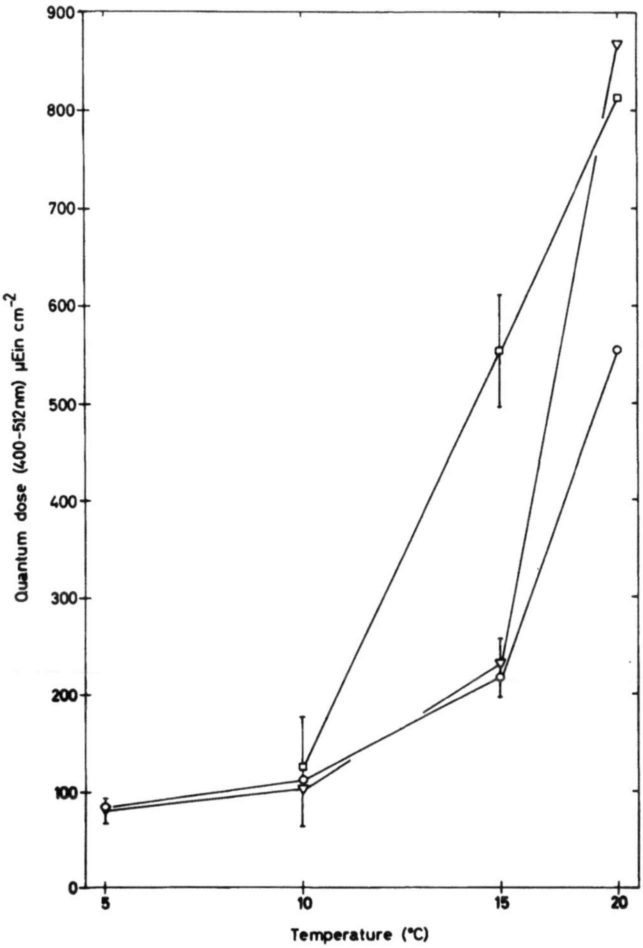

Fig. 2: Quantum dose of blue light (400-512 nm) required to induce fertility in 50% of the female gametophytes of Laminaria hyperborea (circles), L. digitata (squares), and L. saccharina (triangles) at different temperatures. Vertical bars indicate confidence limits at p= 0.05. (Data from ref. 31)

blue light requirement for the induction of gametophyte fertility in the Californian species of Macrocystis pyrifera and Pterygophora californica, as determined at 15 C, is of a similar order to that for L. hyperborea and L. saccharina (38).

The blocking effect of elevated temperatures in regard to induction of fertility, apparent in figure 2 for the temperature range of 10 to 18 C, is complete at 20 - 22 C, which is also the upper range for survival of the various European species of Laminaria (19,31). Only a few distorted sporophytes, or none at all, are formed at these high temperatures (19). Although the basic mechanism of the blue-light mediated morphogenesis is not yet understood, the relationship between the quantum dose of blue light and temperature explains why the common practice of lowering the temperature, without increasing irradiance, serves to induce fertility in laminarian gametophytes.

Figure 3 summarizes the physiological requirements of the gametophytes of three European Laminaria spp. The embryospore, formed by the zoospore, germinates and empties within 24 h to form the primary cell of the gametophyte even in darkness. The primary cell can survive in the dark for at least 5 months. In white fluorescent light the vegetative growth of the primary cell is light-saturated at 20-30 $\mu E\ m^{-2}\ s^{-1}$ (1000-1500 lux). The gametophytes, in the form of the primary cell, are killed at 10 times higher photon flux density and in direct sunlight, at about 350 $\mu E\ m^{-2}\ s^{-1}$, they are killed within minutes. Vegetative growth of the gametophytes increases with temperature and exhibits a broad optimum between 10 and 18 C. The effect of temperature on vegetative growth, therefore, is opposite to that on fertility, since the percentage of fertile gametophytes at a given quantum dose of blue light increases with decreasing temperature. As soon as the primary cell of the gametophyte has increased to a certain critical diameter, due to photosynthesis and growth, the availability of blue light determines whether the pathway to fertility is opened or whether further vegetative growth occurs, such that filamentous thalli are produced. It is also indicated in figure 3 that a spermatozoid-releasing and -attracting substance is released by the female gametophyte at the moment the egg leaves the oogonium, which is a matter of seconds (37).

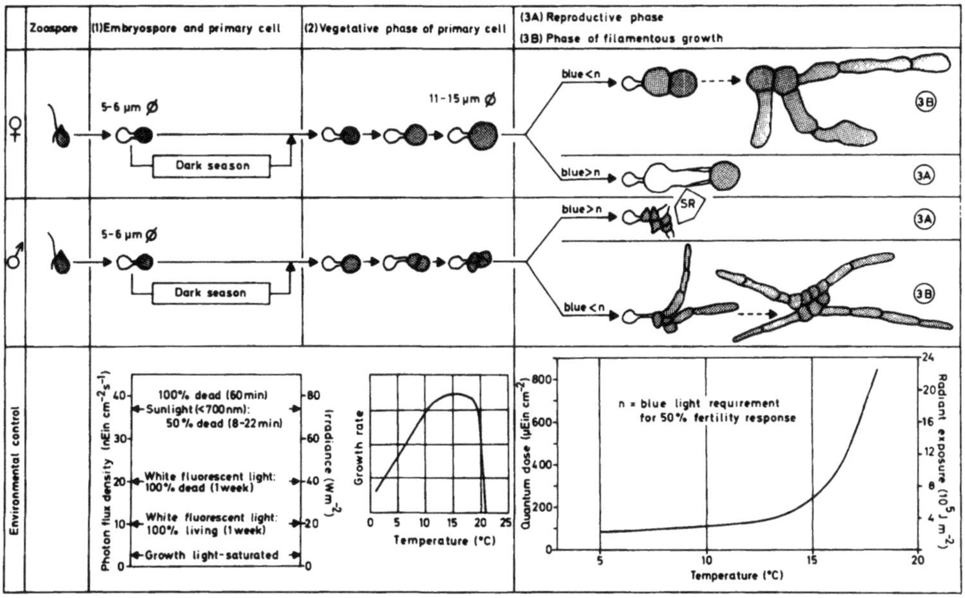

Fig. 3: Scheme summarizing morphological development and physiological requirements of gametophytes of three European species of Laminaria (L. hyperborea, L. digitata, L. saccharina). SR: spermatozoid releasing and attracting substance is released into the water. See text for further explanation. (Data taken from ref. 31)

There is every reason to expect that the availability of blue light in the sea determines the maturation of gametophytes in situ, as it does in the laboratory. The blue light requirement of about 200 µE cm^{-2} at 15 C for the fertility of laminarian gametophytes is similar to that of some other blue light-mediated morphogenetic responses (for examples and references see 35). However, it should be stressed that the magnitude of this blue light requirement is rather high, compared to that of the blue light-mediated photoperiodic response in Scytosiphon lomentaria (35). In S. lomentaria a 50% inhibition of the formation of erect thalli is brought about by 0.002 µE cm^{-2} of blue light, i.e., at a quantum dose 10^5 x smaller than that required for the gametophyte response of the Laminaria spp. at 15 C. To illustrate this difference, it may be said that the photoperiodic response of S. lomentaria is 50% inhibited by giving the alga a night-break of 10 s duration for 10 successive long nights with very low photon flux density (2 µE m^{-2} s^{-1}) blue light (quantum dose per night: 0.002 µE cm^{-2}). In contrast, a 50% response for laminarian gametophytes is achieved by irradiating them with blue light for 28 h at a flux density of 20 µE m^{-2} s^{-1}. If cool white fluorescent light, with 1/5 of the total visible quanta in the blue range, is used, one would obtain a 50% fertility only after 28 x 5 h, or 5.8 days in continuous light, or after 11.6 days if a 12:$\overline{12}$ LD cycle is used.

The ecological significance of these magnitudes has been tested by cultivating laminarian gametophytes in the laboratory in simulated underwater light fields which differ in their blue light content, and also by comparing percentage fertility of gametophytes cultivated in the sea at different water depths where they receive different quantum doses of blue light (31). In a field experiment it was found that the gametophytes which where formed by zoospores of Laminaria saccharina and L. hyperborea in early November survived the dark winter months as unicellular gametophytes due to lack of light even at moderate water depths (36). By the beginning of February, the water suddenly became clear and the experimental gametophytes at 2 m water depth matured within 24 days, the females were all in the unicellular stage, and the males were in a few-celled stage. During these 24 days a quantum dose of blue light of 165 µE cm^{-2} was recorded by an automatic underwater light measuring station at 2 m water depth (31, 36). Water temperature at this time was 5 C. Thus, the quantum dose obtained was sufficient for even a 100% fertility, since in the laboratory experiments a quantum dose of 90 µE cm^{-2} of blue light at 5 C was sufficient for a 50% fertility of the gametophytes (Fig. 2). In another experiment, which also was started from zoospores, but near the end of the sporing season (February, March), gametophytes at 2 m water depth matured within 20 days, the females were again in the unicellular stage. During this period the gametophytes at 2 m water depth had received more than 200 µE cm^{-2} $week^{-1}$. In contrast, the gametophytes at 5 m water depth, which on the average received less than 50 µE cm^{-2} $week^{-1}$ of blue light, formed 2 or 3 vegetative cells in the female or up to 10 cells in the male gametophytes (31). Although the experiment was discontinued at this stage, it is unlikely that multicellular, ultimately macroscopic, gametophytes would have been formed even at 5 m water depth, since the blue light supply at this depth increased considerably during the following weeks and a consequent fertilization of the gametophytes is more likely. In fact, filamentous, macroscopic gametophytes have never been found in nature.

The finding that under sufficient light supply the fertilization of gametophytes in the field occurs in a few weeks corroborates the observations made for Pacific Laminaria saccharina by Hsiao and Druehl (16). They found that the gametophytes of this species from March to May formed oogonia in 12-16 days and antheridia in 10-14 days, and that from November to January the oogonia were formed in 20-26 days and antheridia in 15-18 days.

At Helgoland, where underwater light conditions deteriorate considerably during the winter months - a situation comparable to Arctic conditions in regard to seasonal light supply - there occurs a resting period of the gametophytes in the unicellular stage for 2-3 months, due to a lack of light for photosynthesis. This is followed by a sudden vegetative and reproductive development within 2-3 weeks when light is again available, a situation obtained in early spring near Helgoland and in late summer in Arctic localities (3). Since sufficiently high temperatures prevail during these periods, availability of sufficient irradiance for growth and for induction of fertility seems to be the main factor regulating the seasonal gametophytic development at these locations.

Among other environmental factors that might regulate the seasonal development of the gametophytes, photoperiod has been found not to affect the induction of fertility in the gametophytes of Laminaria digitata, L. hyperborea, L. saccharina, L. japonica, Macrocystis pyrifera, and Chorda filum (32). Although more species should still be investigated to confirm this observation, it may be true that photoperiodism, which is such a characteristic phenomenon among Scytosiphonales (32), does not regulate gametophytic development in the Laminariales. The yearly course of nutrient levels in the sea also does not seem to affect the seasonality of gametophyte development; these gametophytes mature even at very low concentrations of nitrate, provided they are given sufficiently high irradiances in white fluorescent light (Lüning, unpublished). In Pacific

Laminaria saccharina gametophytes, maintained in axenic culture at 10 C and illuminance of 850 lux, fertility was not observed at nutrient concentrations lower than 5 µM NO_3^- and 10 µM PO_4 (15). However, these experiments need to be repeated with higher irradiances.

Temperature may regulate the seasonal development of laminarian gametophytes. If spores are shed in a season in which water temperatures are sufficiently high to block the fertilization of the resulting gametophytes, partially or completely, a seasonal pattern in the occurrence of young sporophytes in the field may be created. An example is provided by *Chorda filum*. The gametophytes of this species mature at 5 and 10 C, but not at 15 C, even at high irradiances in white fluorescent light (32). Near Helgoland the sporophytes appear in April and dissappear by mid September (25). Since the water temperature in May already exceeds 10 C, the gametophytes of this species should be totally blocked from becoming fertile during summer and autumn by elevated temperatures. This may explain why only one generation of macroscopic thalli in *C. filum* is produced per year. Another example of temperature control of fertilization of gametophytes, although not from the order of Laminariales, is provided by *Desmarestia viridis*. This is also an annual alga in which the gametophytes can be fertilized only at temperatures of 5 and 10 C, but not at 15 C (32).

In conclusion, the seasonal development of the gametophytes of *Laminaria* spp., and perhaps also of other genera of the Laminariales seems to be related to the availability of sufficiently high irradiance, and to the reproductive patterns of the sporophytes (see below). It does not seem to be synchronized with a special timing signal, such as a suitable photoperiod; and temperature and nutrient conditions do not seem to be limiting factors under most ecological situations. A similar conclusion was drawn by Hsiao and Druehl (16) in regard to *in situ* gametogenesis of *L. saccharina* and the seasonal appearance of young sporophytes of *Alaria tenuifolia*, *Costaria costata* and *L. saccharina* from British Columbia (9).

Phenological Patterns in Sporophytes of *Laminaria* spp.

The seasonal events in sporophyte performance of a few perennial representatives of the Laminariales, for which detailed information is available about vegetative growth as well as reproductive activity, will be exemplified in the following section by the phenological patterns of four *Laminaria* spp. (Table I).

In *Laminaria hyperborea* vegetative growth of the blade is strictly limited to the period from December to June, the fastest expansion in area occurring in April/early May (22,23,27,28). Also, stipe elongation (22,27) as well as an increase in stipe diameter (17) are fastest during this period although growth activity in these two parameters continues at a slower rate during the second half of the year. The reproductive activity of *L. hyperborea* sporophytes is also confined to a limited

Table I. *Laminaria* spp.: Seasonal patterns of vegetative growth and reproduction in sporophytes. V: vegetative growth. S: sori present on blades. Underlined letters: seasonal peak. Letters in brackets: valid for only part of the month. Dash: no activity.

Species	Locality	J	F	M	A	M	J	J	A	S	O	N	D	Ref
L. hyperborea	Isle of Man	S̲	S	S	(S)	-	-	-	-	S	S	S	S	(21)
	Helgoland	V	V	V̲	V̲	V	(V)	-	-	-	-	-	(V)	(27,28)
		S	S	S	(S)	-	-	-	-	-	-	S	S	(*)
L. digitata	Calvados	V	V	V	V̲	V̲	V̲	V	V	V	V	V	V	(5,40)
		S	S	S	S	S	S	S̲	S̲	S̲	S̲	S	S	(6)
	Wales	S	S	-	-	-	-	-	-	S	S	S	S̲	(42)
	Helgoland	-	-	-	-	S	S	S	S	S	S	S	(S)	(*)
L. saccharina	Devon and	V	V̲	V̲	V̲	V̲	V̲	V	V	V	V	V	V	(39)
	Argyll	S̲	S̲	S̲	S	S	S	S	S	S	S̲	S̲	S̲	(39)
	Wales	S̲	S̲	S	(S)	-	-	-	-	-	(S)	S	S	(42)
	Helgoland	S	S	(S)	-	-	-	-	-	(S)	S	S	S	(*)
L. angustata	Hokkaido	V	V	V	V	V̲	V̲	V̲	V	V	V	V	V	(13)
		S	-	-	-	-	S	S	S	S	S̲	S	S	(13)

(*) - This paper.

period -- from September to early April on the Isle of Man and from November to early April at Helgoland (Table I). L. hyperborea is thus a winter sporing species, although it should be emphasized that the appearance of sorus precedes the appearance of new blade by up to 3 months and reproductive activity is largely finished when the maximum vegetative growth occurs (Table I). In summer only vegetative sporophytes are present. The species seems to react in a similar way throughout its whole geographical range (21), from Portugal up to the North Cape of Norway (20,24).

In contrast, blade growth in L. digitata and L. saccharina is continuous during the year, but maximum elongation rates occur, as in L. hyperborea, during the first part of the year (Table I). At Helgoland, L. digitata is the earliest of all three Laminaria spp. to produce sori in the course of the year. The first sporogenous plants from which one may obtain zoospore release appear by May, whereas in L. saccharina they do not appear until mid September. The last sporogenous plants of L. digitata at Helgoland are seen up to December and of L. saccharina up to March of the next year. Thus, in these two species also, as in L. hyperborea, the reproductive activity is stopped partly or completely when rapid vegetative growth takes place in early spring. A somewhat similar reproductive pattern has been recorded for L. digitata and L. saccharina from Wales (42); for L. saccharina, no sporogenous plants can usually be found on the Port Erin breakwater in summer (21). L. digitata has also been reported to behave as a summer and autumn sporing species along the coast of Calvados (France), with a maximum number of sporogenous plants and maximum spore emission occurring from June to October (6; for discussion of conflicting reports, see 11). However, at the locality cited, sporogenous plants also occur during the rest of the year, but at a lower frequency. The same case, with sporogenous plants all year round, has been reported for L. saccharina from France (45), and from Devon and Argyll (British Isles) by Parke (39). These findings are somewhat at variance with those of Rees (42) and Kain (21); however, this variance may be due to the existence of local, genetically different populations. L. saccharina from Helgoland, in which no sporogenous plants can be found in the field from April to August, definitely represents a genetically different form, with smooth blades, as opposed to the form with rugose blades otherwise common in Europe (29, 33).

In Laminaria angustata from Hokkaido (Japan) the maximum vegetative growth occurs from May to July, as reported by Hasegawa (13). The plant first becomes sporogenous in October, at a minimum age of 8 months. In the second year the sporogenous activity begins in July, reaches a maximum in October and ceases in the following February. In its third year the plant is sporogenous from October to the next January, again with a peak in October (data in Table I refer to third year). The seasonal patterns of vegetative growth and reproduction are thus similar to the European L. saccharina.

The general picture, then, is that the adult sporophytes of Laminaria spp. considered here have a seasonal peak of vegetative growth in spring and a reproductive maximum in autumn or winter, the exact time period depends on the species and perhaps also on location. Vegetative growth is minimal when maximum reproduction occurs. After the seasonal peak of vegetative growth the fronds require a period of several months to form mature sori. In L. hyperborea this period lasts for at least 3 months (Isle of Man, June to August, Table I) or 5 months (Helgoland, June to October, Table I). In addition, it is known that L. hyperborea sporophytes must have a minimum age of 15 months to become sporogenous (21). For L. saccharina from the British Isles, Parke (39) reported that sori are not formed on blade portions younger than 6 months, and for Helgoland L. saccharina Lüning (30) reported that the corresponding age may be 4 months. "Resting periods" of similar magnitude may be valid for L. digitata, and the minimum age of a sporophyte of this species to become sporogenous has been reported to be 18-20 months (41).

New sporophytes of the three European species of Laminaria, of a size just visible to the naked eye, appear in the field in most localities mainly in the winter and early spring. This is to be expected from the sporing periodicity indicated in Table I. Using sterilized stones of about 10 kg, which were exposed for two-month periods in the upper sublittoral in the Isle of Man, Kain (21) found that juvenile sporophytes of Laminaria spp. were formed in great quantity on stones exposed during November/December and January/February, a few were also found on stones exposed during March/April, but none on stones exposed during the rest of the year. The largest L. hyperborea sporophyte which developed in the sea at moderate water depth in October/November after 33 days consisted of 167 cells (22), equivalent to a length of about 0.2 mm (18). In another field experiment, which was conducted in May, Kain (18) found that only 25 days from spore release were required for the formation of 0.6 mm long sporophytes (1000 cells); from this she calculated a total of 54 days for the formation of 1 cm long sporophytes. These data are consistent with those from laboratory experiments in continuous light, in which 2-3 cm long sporophytes of L. hyperborea, L. digitata and L. saccharina have been grown from gametophytes within 2 months (30). These results indicate that young sporophytes of a size just visible to the naked eye would appear in the field with a lag phase of several months after the seasonal reproductive peak of the sporophytes.

Possible Role of Environmental Factors in Regulation of Sporophyte Periodicity

For those laminarian species which show a clear seasonal pattern of sorus formation and for those in which production of the new blade commences abruptly, as in Laminaria hyperborea in December, the question arises: Does the plant receive an environmental signal and, if so, what is its nature? For sorus formation in L. saccharina, Parke (39) assumed that "the development of the reproductive tissue in plants of this species is controlled by some factor or factors within the frond tissue of the plant, not by factors external to the plant." Lobban (26) suggested that in Macrocystis spp. the apical meristem might "leave some inhibitor or repressor in the laminae that are to remain sterile." It is possible, therefore, that for frond tissue in Laminaria spp. the distance from the meristem, and thus the age of a particular frond portion, may be an important precondition for onset of sorus formation.

However, a completely internal control does not seem very likely for Laminaria hyperborea , in which plants of all age groups (maximum possible age: 15-18 years; 24) stop growth by early summer, become sporogenous from autumn onwards and commence new growth by December (Table I). Internal, annually cycling rhythms of activity would be badly put out of step due to the innate variability of living things with time, as pointed out by Villiers (48). If some environmental factor induces and synchronizes kelp performance with the seasons, then, at first glance temperature or nutrient concentration or photoperiod might all represent such a factor. However, so far our knowledge in this respect is almost nil, and, hence, in the following, I will attempt to pull together the few fragments of information which may suggest directions for future research.

In a few cases, such as Alaria esculenta (47), plants have been reported to become sporogenous following their transfer from the sea in a vegetative stage. However, in these cases, the plants may have been induced to become sporogenous while still in the sea before their transfer to the laboratory. Only rarely has it been possible to cultivate the sporophytes of Laminariales completely to the stage of sorus production. Sanbonsuga and Hasegawa (43; see also 14) achieved sorus production in Undaria pinnatifida (length of sporogenous plants: 15-60 cm) and in Costaria costata (maximum length of sporophytes obtained: 50 cm; sorus formation after a certain decrease in length), both of which are annual plants. Sori were formed in these two species at temperatures above 22 C, at an illuminance of 2500 lux and at 18 h light per day. In the field, at Hokkaido, both species begin to form sori from early summer onwards and attain full maturity by mid summer. In additional experiments, Sanbonsuga and Hasegawa (44) found that temperature (10-12 C or 18-20 C), illuminance (2500 or 5000 lux), and photoperiod (6 or 18 h light per day) are unlikely to represent triggering factors for sorus formation in Costaria costata, since sorus formation occurred in laboratory-reared sporophytes of this species under all conditions stated. These authors made the interesting observation that a period of retarded frond growth always preceded the onset of sorus formation, and that "the retardation of frond growth and onset of zoosporangium formation could be induced in mature plants, but not in the immature, by decreasing nourishment in the culture medium." Low nitrate content is known for several algae to be a precondition for onset of reproduction (see review by Dring, 8), and has, in addition, definitely been shown to be responsible for slow growth rates of Laminaria longicruris during summer off the coast of Nova Scotia (2). The observation of Sanbonsuga and Hasegawa (44) on induction of sorus formation by low nutrient content should be taken as a possible starting point for further analysis of the problem.

However, as shown below, under identical environmental conditions the different Laminaria spp. react in a species-specific way, and it is hard to believe that this could be due to a species-specific nutrient requirement. This is evident from figure 4 which summarizes an experiment performed by Lüning (30). Laboratory-raised sporophytes of L. digitata, L. saccharina, and L. hyperborea were transferred by mid February (blade lengths: 5-25 cm) into the sea near Helgoland and their growth was followed until November. The experimental plants of L. hyperborea stopped growth completely in July, L. saccharina reduced its growth rate considerably during this month, wheras L. digitata reduced its growth rate first in September. The most puzzling aspect in regard to these species-specific patterns of seasonal growth reduction was the abrupt stop in growth of the L. hyperborea plants in July. Water temperature in this month surpassed 15 C, but the plants did not resume growth when it fell below 15 C in October. Irradiance was still saturating for photosynthesis in August, and nitrate concentration was still as high as 8 μM NO_3^-, near enough to the growth-saturating concentration of 10 μM NO_3^- reported for L. saccharina sporophytes by Chapman et al. (4). Reduced summer growth under cold, NO_3^- - enriched conditions has also been reported for Alaria esculenta from Newfoundland by Buggeln (1), who speculated whether an endogenous, true circannual rhythm may not be involved in bringing about this reduced growth in summer.

In many higher plants of the temperate region, a suspension of growth and induction of dormancy is common after the plants have recorded a signal from the environment, namely a decrease in daylength, which occurs from late June onwards in the Northern hemisphere (48,49). This photoperiodic signal represents an "early warning system" and occurs, when growth conditions, such as temperature, irradiance and nutrient supply are still favourable. In this context, it should be clarified that plants which

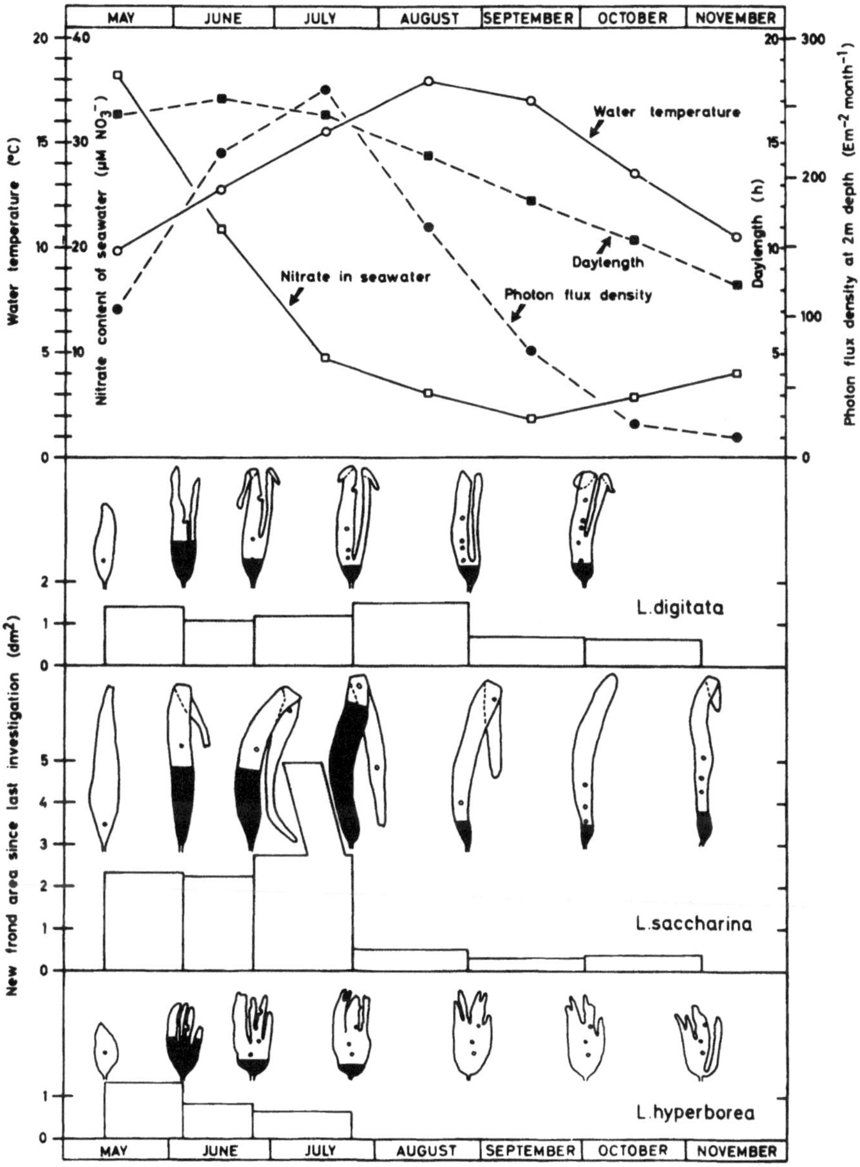

Fig. 4: Helgoland, 1975. Seasonal course of temperature, nitrate concentration in seawater (from ref. 12), photon flux density at 2 m water depth (from ref. 36), daylength (according to Smithsonian Meteorological Tables), and of frond area increase in experimental sporophytes of three Laminaria spp. (bar graphs), cultivated at 2 m water depth. Filled area in outline drawings: new-formed tissue. (Data taken from ref. 30)

reduce their growth rate due to the lack of some factor, such as nutrients, and resume their growth when this lack has been rectified, are by definition only in a "quiescent state" (48). The Nova Scotian Laminaria longicruris, which is reported to resume growth in summer after artificial increase of nitrate content in the sea water (2), may be an example showing quiescence. In contrast, the ceassation of growth by L. hyperborea in summer may represent true "dormancy," which can only be broken by another environmental signal. In higher plants, this may be photoperiod, or, more often, a minimum period of temperatures below 5 C (48). Some of the photoperiodic responses reported in marine algae may be interpreted in these terms. For instance the swarmers of the erect thalli of the brown alga Scytosiphon lomentaria form minute prostrate systems, which do not form any new fronds, in long days (32). The critical daylength for this reaction increases toward the North Pole in the latitudinal ecotypes of this species. The ecological significance of the response is particulary evident in subarctic S. lomentaria, in which a small percentage of the swarmers still form erect fronds at 16 h light per day, but not at 18 h light per day or in continuous light, so that by midsummer only prostrate crusts are formed, which probably have a better chance of survival under ice than a new generation of erect fronds.

The photoperiodic induction in Scytosiphon lomentaria occurs over a broad temperature range. This is not so for the photoperiodic induction of tetrasporangium formation by short days in the Trailliella-phase of the red alga Bonnemaisonia hamifera (32), which occurs over a narrow temperature range around 15 C and requires, in addition, low nutrient concentrations. Thus, it may be that the temperature and nutrient content of the water are additional factors and limit the range of possible photoperiodic induction to a certain set of environmental conditions. For obvious reasons, photoperiodism in marine algae has been searched for and discovered so far mainly in smaller forms (32; see also reviews by Dring, 7,8). Nevertheless, it seems more important now to ascertain whether or not photoperiodic reactions are also involved in the control of seasonal patterns of kelp sporophytes.

References

1. Buggeln, R.G.: J. Phycol. 14, 156-160 (1978).

2. Chapman, A.R.O., Craigie J.S.: Mar. Biol. 40, 197-205 (1977).

3. Chapman, A.R.O., Lindley, J.E.: Mar. Biol. 57, 1-5 (1980).

4. Chapman, A.R.O., Markham, J.W., Lüning, K.: J. Phycol. 14, 195-198 (1978).

5. Cosson, J.: Bull. Soc. linn. Normandie 8, 246-281 (1967).

6. Cosson, J.: Soc. Phycol. de France 21, 28-34 (1976).

7. Dring, M.J.: In Photobiology of Microorganisms (P. Halldal, ed.), pp. 345-368, Wiley & Sons, London 1970

8. Dring, M.J.: In Algal Physiology and Biochemistry (W.D.P. Stewart, ed.), pp. 814-837, Blackwell, Oxford 1974

9. Dring, M.J., Lüning K.: Planta (Berl.) 125, 25-32 (1975).

10. Druehl, L.D., Hsiao, S.I.C.: J. Fish. Res. Board Canada 34, 1207-1211 (1977).

11. Gayral, P., Cosson, J.: Synop. FAO Pêches 89, pp. var. (1973).

12. Harms, E., Hagmeier, E.: In B.A.H. Jahresberichte 1975, pp. 21-27, Westholsteinische Verlagsdrückerei Boysens & Co, Heide, F.R.G. 1976

13. Hasegawa, Y.: Bull. Hokkaido reg. Fish. Res. Lab. 24, 116-138 (1972).

14. Hasegawa, Y., Sanbonsuga, Y.: In Contributions to the Systematics of Benthic Marine Algae of the Northern Pacific (I.A. Abbott, M. Kurogi, eds.), pp. 109-116, Japanese Society of Phycology, Kobe, Japan 1972

15. Hsiao, S.I.C., Druehl, L.D.: Can. J. Bot. 51, 829-839 (1973).

16. Hsiao, S.I.C., Druehl, L.D.: J. Phycol. 9, 160-164 (1973).

17. Kain, J.M.: J. mar. biol. Ass. UK 43, 129-151 (1963).

18. Kain, J.M.: J. mar. biol. Ass. UK 45, 129-143 (1965).

19. Kain, J.M.: J. mar. biol. Ass. UK 49, 455-473 (1969).

20. Kain, J.M.: FAO Fish. Synop. 87 65 pp. (1971).

21. Kain, J.M.: J. mar. biol. Ass. UK 55, 567-582 (1975).

22. Kain, J.M.: J. mar. biol. Ass. UK 56, 267-290 (1976).

23. Kain, J.M.: J. mar. biol. Ass. UK 56, 603-628 (1976).

24. Kain, J.M.: Mar. Biol. Ann. Rev. 17, 101-161 (1979).

25. Kornmann, P., Sahling, P.-H.: Helgoländer wiss. Meeresunters 29, 1-289 (1977).

26. Lobban, C.S.: Phycologia 17, 196-212 (1978).

27. Lüning, K.: Mar. Biol. 2, 218-223 (1969).

28. Lüning, K.: In Fourth European Marine Biology Symposium (D.J. Crisp,

ed.), pp. 347-361, University Press, Cambridge 1971

29. Lüning, K.: Helgoländer wiss. Meeresunters. 27, 108-114 (1975).

30. Lüning, K.: Mar. Ecol. Prog. Ser. 1, 195-207 (1979).

31. Lüning, K.: J. Phycol. 16, 1-15 (1980).

32. Lüning, K.: In The Shore Environment, Vol. 2. Ecosystems (J.H. Price, D.E. Glrrine, W.F. Farnham, eds.), pp. 915-945, Academic Press, London 1980

33. Lüning, K., Chapman, A.R.O., Mann, K.H.: Phycologia 17, 293-298 (1978).

34. Lüning, K., Dring, M.J.: Planta (Berl.) 104, 252-256 (1972).

35. Lüning, K., Dring, M.J.: Mar. Biol. 29, 195-200 (1975).

36. Lüning, K., Dring, M.J.: Helgoländer wiss. Meeresunters. 32, 403-424 (1979).

37. Lüning, K., Müller, D.G.: Z. Pflanzenphysiol. 89, 333-341 (1978).

38. Lüning, K., Neushul, M.: Mar. Biol. 45, 297-309 (1978).

39. Parke, M.: J. mar. biol. Ass. UK 27, 651-709 (1948).

40. Pérez, R.: Proc. Int. Seaweed Symp. 6, 329-344 (1969).

41. Pérez, R.: Rev. Trav. Inst. Pêches Marit. 35, 287-346 (1971).

42. Rees, T.K.: Proc. Swansea Sci. Fld. Nat. Soc. 1, 33-37 (1928).

43. Sanbonsuga, Y., Hasegawa, Y.: Bull. Hokkaido, reg. Fish. Lab. 32, 41-48 (1967).

44. Sanbonsuga, Y., Hasegawa, Y.: Bull. Hokkaido, reg. Fish. Lab. 35, 198-202 (1969).

45. Sauvageau, C.: Mém. Acad. Sci. Inst. Fr. 56, 1-240 (1918).

46. Schreiber, E.: Planta 12, 331-353 (1930).

47. South, G.R.: Helgoländer wiss. Meeresunters. 20, 216-228 (1970).

48. Villiers, T.A.: Dormancy and the Survival of Plants, Edward Arnold, London 1975

49. Vince-Prue, D.: Photoperiodism in Plants, McGraw Hill, London 1975

50. Yabu, H.: Mem. Fac. Fish. Hokkaido Univ. 12, 1-72 (1964).

DISCUSSION

FOSTER: I was curious. Having kept these gametophytes for 5 months in the sea, was there not enough cumulative blue light to set those off over that period?

LÜNING: In winter our storms begin in October and then the water is practically brown until February. That is the normal situation. From year to year, there may be a phase of two weeks when suddenly the storms cease and all this dirt falls down, then certainly there might be fertilizations. As I said we have an Arctic condition, we have dirt instead of ice, therefore, I gave the example from the Isle of Man, where we have sporophytes from November, December onwards as you would expect.

DRUEHL: We have used the blue light, or Lüning effect, very successfully. There is one species, however, that totally fails to respond, that is, Postelsia. It is strictly an intertidal plant, and obviously experiences very different light regimes. I was wondering if you could comment on this?

LÜNING: I have a whole collection of laminarian gametophytes, 40-50 species, and there are certainly 15 or so that we can not fertilize at all. So there is more than just blue light effect. These are just first observations. I should also say that it is improbable that such a photomorphogenetic pigment exists in only one form. All such pigments which have been found -- phytochrome, rhodopsin, etc. -- exist in two forms and now, for the last two years or so, the photobiologists are discovering that blue light effect in higher plants may also be blue-yellow effects. We are nowhere near the end of these investigations, and I do not know what happens and what other substances may be there.

DUNCAN: You talked about the single female cells. Were you talking about Laminaria hyperborea, and were these in plastic dishes?

LÜNING: All three Laminaria species did the same. They were growing on glass slides, but in Plexiglas containers filled with Provasoli. Provasoli has about 300 micromolar nitrate, but we know that even at 1 micromolar nitrate, if you always renew the medium, you can fertilize them.

DUNCAN: You never got multicellular females?

LÜNING: No. Well, I got them at depths of six metres, but I did not continue there. We need more experiments. When I was in Santa Barbara, we put Macrocystis down and saw some multicellular gametophytes, but again some of the cells had formed eggs.

WILLENBRINK: How specific are the attractants in the case of the three Laminaria species?

LÜNING: That is easy. Everybody releases everybody, there. Macrocystis, Laminaria hyperborea -- we mix them all.

WILLENBRINK: But not the fucoserraten.

LÜNING: Our Laminaria do not really extract Fucus spermatozoids. As you may know, there are now five classes of sex attractants in the plant kingdom. The first one, found in 1968 was found in a fungus. The remaining four have all been from brown algae (EDITOR'S note: Here Dr. Lüning reviewed the interesting work of Dr. D.G. Müller of Konstanz, F.R.G. and the properties of ectocarpen, multifiden, fucoserraten, etc. For an introductory reference to that work see Science 212, 1040 [1981]). In the Laminariales it seems that Macrocystis, Laminaria, Alaria, and so on, all produce the same substance. On the gas chromatogram it is always the same peak. But there are tremendous difficulties in trying to concentrate the substance for further characterization. Laminaria is the only case, so far, where this gas also releases the spermatozoids.

FLOC'H: Une question concernant la croissance de Laminaria hyperborea. Es-ce que le commencement est toujours en Decembrè, jamais avant?

LÜNING: Always Decembrè.

SRIVASTAVA: A small question about the blue light effect. I am trying to figure out whether it is a morphogenetic effect in the sense that the sperms or eggs are actually formed after blue light is shone or whether it is simply a membrane effect - that they are formed earlier, and simply released by shining of blue light.

LÜNING: No. It is a complete change in the program. The formation of the oogonium is controlled by the blue light, but the release is controlled by red light. We do not know much more about it.

SRIVASTAVA: From what you are saying it seems to me more likely that it probably is some kind of a hormone release which can trigger the morphogenetic process.

LÜNING: Yes. That is possible. But the hormone, of course, is only a secondary cause.

DRUEHL: Considering the general nature of this sex attractant in *Laminaria*, one would expect to see a great deal of hybridization in nature in *Laminaria*.

LÜNING: I do not think that is a problem. Attracting a spermatozoid does not mean much. It is the intrusion of the spermatozoid and acceptance of the nucleus, that is where all the blocks begin. It seems to be a principle of nature to have this rather unspecific release as we have seen already in *Fucus serratus* and *F. vesiculosus*, probably also in others. Now if there is a population of *Fucus serratus*, there is normally only *Fucus serratus* there, and these things certainly act only over a very short distance. Therefore, there is not much danger that hybridization would occur. Müller has put it the other way around. He said, " maybe the fact that *F. serratus* produces the same attracting substance as *F. vesiculosus* is like a kind of chemical warfare, to attract the others."

SCHMITZ: Are there any analogs to the sex substance in the Laminariaceae which could be applied, perhaps to vegetative cells, to elicit the same response?

LÜNING: No, not yet. They are still trying to establish the structure.

Part II. Nutrition and culture

INORGANIC NUTRITION OF MARINE MACRO-ALGAE IN CULTURE[1]

J. McLachlan
Atlantic Research Laboratory, National Research Council of Canada,
1411 Oxford St., Halifax, N.S., Canada B3H 3Z1

Introduction

Until about 25 years ago, relatively few species of marine algae, including both unicells and macrophytes, were being cultured. Culturing of species in the latter category was in many instances limited to spore germination and perhaps several subsequent cell divisions. Improved marine media became available in the latter part of the 1950's (64) resulting in a dramatic increase not only in the variety of species available in culture but also in the number of workers using algal cultures as research tools. Bold in 1974 (10) concluded that all major advancements in phycology during the past several decades were attributable to the availability and use of algal cultures. Certainly, even a casual perusal of current literature verifies that algal cultures are used in a large variety of research projects, although the majority of investigations have utilized unicellular species.

It was not until the immediate past decade that attention was focused on the culturing of marine macro-algae. Since then, there has been a considerable increase in the variety of investigations undertaken, although compared with unicellular species, work with marine macro-algae still remains limited and information on their nutritional requirements is meagre. I believe this to be the result of two factors, probably the most important being the absence of any established culture collections for macrophytes. This is in spite of the obvious need for cultures, as emphasized by Bold (10), and repeated pleas for the establishment of collections (62,63,68). Consequently, studies that have been done

[1] NRCC No. 18248

Synthetic and Degradative Processes in Marine Macrophytes

generally reflect the interests of those who have isolated the organisms. Maintenance of macro-algae in culture perhaps requires more time and space than that required for unicellular species, but even so the situation with micro-algae is becoming desperate ("Phycological Newsletter", vol. 15, No. 3, 1979). Secondly, only a few macro-algae have been isolated into axenic culture. Many of the studies undertaken have not required axenic cultures; moreover, investigators and reviewers have not demanded them, but at one time it was impossible to publish on 'bacterized' unicellular species. As a result of this paucity of axenic cultures of marine macrophytes, there is only limited information on organic nutrition, and this is fragmentary and sometimes contradictory. I will not consider organic nutrition as this aspect has been reviewed by Fries (24). I must emphasise, though, that this field is wide open, and because of developmental processes associated with macrophytes, many exciting and challenging problems await phycologists. Information on inorganic nutritional requirements is equally scanty, partly because macro-algae have generally been cultured in media previously designed for culture of unicellular species. Several of these media have been generally successful, thus obviating the necessity for designing new or special media. Moreover, there has been a general lack of employment of artificial media, and most have been singularly disappointing. Because of the limited variety of studies on macrophytes in culture, it is difficult to establish general principles. Instead I can only point to a few interesting phenomena, and suggest areas where deficiencies in information urgently require attention.

Essential Elements

Nutritional studies must be done with cultures, and of course absolute requirements can only be established through culture studies. Criteria for an absolute requirement were established by Arnon (4), based on the postulate that deficiency of a nutrient renders it impossible for the alga to grow, to photosynthesise, or to complete the vegetative and reproductive stages in its life history. The deficiency must be specific, i.e. only that nutrient can prevent or correct the deficiency. Moreover, the nutrient must be involved directly in the nutrition of the plant and not related to a chemical or biological condition of the medium (4). The latter condition is especially difficult to resolve because of the balance and interaction of elements in the concentrated and complex marine type of media. Only in a very few instances have absolute requirements been established for marine algae.

Table I: Essential elements in plant nutrition

Macronutrients		Micronutrients	
Metals	Non-Metals	Metals	Non-Metals
Potassium	Carbon	Iron	Boron
Sodium	Hydrogen	Copper	Silicon
Calcium	Oxygen	Zinc	Chlorine
Magnesium	Phosphorus	Molybdenum	Iodine
	Nitrogen	Cobalt	Bromine
	Sulphur	Vanadium	

More than 100 years ago, both Knop and Sachs formulated media for growth of higher plants in hydroponic culture (22), thereby identifying the essential major nutrients. Molish, nearly 80 years ago, noted that the mineral nutrition of algae was not unlike that of higher plants (64), a conclusion that has been repeatedly substantiated. Indeed, Knop's medium is still commonly used, not only for higher plants but also for various other plants including freshwater algae. Sachs included NaCl in his medium and these two elements are now considered essential for plants (Table I). It is interesting that sodium and chlorine are the predominant elements in seawater (Table II), and, hence, in marine media, whereas nitrogen and usually calcium dominate media for freshwater algae and terrestrial plants.

Those elements considered essential for plants are listed in Table I, and all of these have been identified as being required by at least one algal species (22, 59). Bromine is now regarded as essential for several marine macro-algae (25,50) and must be added to the list of required elements for plants. It is unnecessary to emphasise that if an absolute requirement

Table II: Major elements in seawater of 35 $^{o}/oo$ salinity (after Collier [16]).

Element	g/kg	mmol/kg
Cations		
*Sodium	10.556	459.02
*Potassium	0.380	9.72
*Magnesium	1.272	52.30
*Calcium	0.400	9.98
Strontium	0.0085	0.15
Anions		
*Chlorine	18.980	535.30
*Bromine	0.065	0.81
Fluorine	0.001	0.05
*Sulfate	2.649	27.57
*Bicarbonate	0.140	2.29
*Boric acid	0.026	0.42

* Essential for plants.

is demonstrable then that nutrient must be present in the environment of the plant. But it is important to appreciate that the nutrient must be in an assimilative form, an important consideration with marine media. As is to be expected, all elements known to be essential for plants occur in seawater (Tables II, III). Moreover, the major constituents of this medium, except for strontium and fluorine, are required by plants, and other than chlorine are macronutrients (Table I). Seawater contains numerous other elements; except for those present in very minute concentrations, these are listed in Table III. Unlike the major elements, only a few of these minor elements are known to be essential for plants.

The essential elements are accumulated in the tissues of algae over their concentration in seawater (Table IV, cf. with Tables II and III). The concentration factor of some of the trace elements may be many thousand fold. The elemental composition of the ash of macro-algae is similar to that of phytoplankton, and interestingly, the concentration of elements considered adequate for higher plants (22) is similar to that found in marine algae.

Table III. Minor (trace) elements[a] in seawater of 35°/oo salinity (after Goldberg [29]).

Element	mg/l	Element	mg/l
*Si	3.0	Cs	0.0005
A	0.6	Se	0.0004
*N	0.5	La	0.0003
Li	0.17	Y	0.0003
Rb	0.12	Kr	0.0003
*P	0.07	Cd	0.00011
*I	0.06	X	0.0001
Ba	0.03	W	0.0001
In	<0.02	*Co	0.0001
*Fe	0.01	Ne	0.0001
Al	0.01	Ge	0.00006
*Zn	0.01	Cr	0.00005
*Mo	0.01	Th	0.00005
As	0.003	Ag	0.00004
*Cu	0.003	Sc	0.00004
U	0.003	*Ga	0.00003
*Mn	0.002	Hg	0.00003
Ni	0.002	Pb	0.00003
*V	0.002	Bi	0.00002
Ti	0.001	Nb	0.00001
Sn	0.0008	Tl	<0.00001
Sb	0.0005		

[a] Other elements are present, but in concentrations $<10^{-5}$ mg l^{-1}.

* Essential for plants.

Many, perhaps all, elements in seawater can be assimilated by the algae. Nevertheless, neither the presence of an element in the ash of a plant nor correlated increase in biomass with decreased concentration in the medium, as for instance has been shown for silicon (27) and aluminum (47,69), are indicative of a requirement or even that the element in question has been assimilated.

Table IV. Composition of Ash

Element	Chondrus crispus (83)	Laminaria digitata (7)	Nereocystis luetkeana (78-80)	Ulva lactuca (46)	Phyto-Plankton (11)	Adequate for higher plants (22)
% dry wt						
*Na	3.6	-	4.4	3.5	10.6	-
*K	3.4	-	19	2.5	1.3	1.0
*Ca	1.3	-	0.6	0.9	0.53	0.5
*Mg	-	-	0.6	-	1.5	0.2
*S	-	-	0.5	-	-	0.1
*Si	-	-	0.5	-	7.9	-
ppm						
F	8.3	-	-	6.9	-	-
*Mo	0.6	0.2	-	0.5	-	0.1
*Co	0.4	0.3	<0.2	0.7	-	-
*Cu	20	5	<2	62	7.4	6
As	5	-	-	4	-	-
Ni	>2.0	1.8	-	>2	7.8	-
*Mn	-	<30	<4	-	13.3	50
*Zn	64	59	5	62	122	20
*I	271	-	1163	50	-	-
*Fe	3	400	38	-	1510	100
Pb	-	7	0.6	-	7.2	-
Sn	-	0.6	-	-	-	-
V	-	0.5	-	-	-	-
Ti	-	4	-	-	27	-
Cr	-	1.1	<2	-	3.9	-
Ag	-	0.1	-	-	-	-
Sr	131	690	668	-	119	-
Ba	56	17	5	-	19	-
Al	-	-	49	-	444	-
*Br	-	-	<79	-	-	-
*B	-	-	89	-	-	20
Ag	-	-	-	-	0.6	-
Cd	-	-	-	-	1.6	-
Hg	-	-	-	-	0.16	-

* Essential for plants.
- Not determined.

Culture Media

As all elements essential for plants occur in seawater, it follows that this medium should support good growth of algae and other plants, and indeed this is the case. Unfortified seawater has been used in experimental work for macro-algae (e.g., 28), but rapidly becomes deficient (50), a fact apparent early in the history of culturing marine algae (35,64). The problem is that some of the required major elements are present in seawater as minor constituents, most noticeably nitrogen and phosphorus, and to a lesser extent silicon (Table III). In most habitats, benthic species are washed continuously by large volumes of water providing a steady, albeit dilute, supply of nutrients. Studies of substrate concentrations, especially in flow-through systems, are thus important (e.g., 13). During part of the year, however, the concentration of some nutrients, in particular nitrogen, may fall below the limits of analytical detection. Strategies for accumulation of essential elements have been evolved in some species of macro-algae that permit accommodation to periods of nutrient deficiency in the environment (53).

Supplementation of seawater with nitrogen and phosphorus, as done by Schreiber (65), provides a satisfactory medium for growth of many algae (49), including macrophytes. Soil extract may also be added. In some cases iron may replace soil extract in the Erd-Schreiber preparation (49). In actively growing cultures, requirements for nitrogen and usually phosphorus are easily demonstrable (e.g. 50); the same applies to species requiring major amounts of silicon. Although the Schreiber (or Erd-Schreiber)-type of medium may still be used, media now more commonly employed contain additions of other nutrients, both inorganic and organic. The inclusion of trace elements and synthetic chelators (64) is unquestionably the most significant improvement over the Schreiber-type of medium, and has resulted in the culturing of a large number and a broad spectrum of species. These fortifications are made to enriched seawater (30,31,49,62,63) as well as artificial seawater media (49,64). The composition of two of these media, SWM and ASP-M, which we at the Atlantic Research Laboratory have used, is given in Table V.

Table V. Composition of SWM (enriched seawater) and ASP-M (artificial) media for culture of macrophytes (modified after McLachlan [49])

Additive	SWM	ASP-M
NaCl	–	400 mM
KCl	–	10 mM
$MgSO_4$	–	20 mM
$MgCl_2$	–	20 mM
$CaCl_2$	–	10 mM
$NaHCO_3$	–	2 mM
$NaNO_3$	0.1–0.3 mM	0.1–0.3 mM
NaH_2PO_4	10 µM	10 µM
Na_2SiO_3	0–0.2 mM	0–0.2 mM
EDTA	27 µM	27 µM
Fe	2 µM	2 µM
Cu	0.2 nM	0.2 nM
Co	0.2 µM	0.2 µM
Mo	5 µM	5 µM
Mn	7 µM	7 µM
Zn	0.8 µM	0.8 µM
B	–	0.4 mM
Br	–	0.5 mM
Sr	–	0.1 mM
Rb	–	2 µM
Li	–	10 µM
I	–	0.2 µM
Vitamin mix	a	a
TRIS	0–5 mM	0–5 mM
pH	7.5	7.5

[a] Conc. per litre of medium: 0.5 mg Thiamine·HCl, 0.1 mg Nicotinic acid, 0.1 mg Calcium pantothenate, 10 µg p-Aminobenzoic acid, 1.0 µg Biotin, 5.0 mg i-Inositol, 2.0 µg Folic acid, 1.0 µg Cyanobalamin, 3.0 mg Thymine.

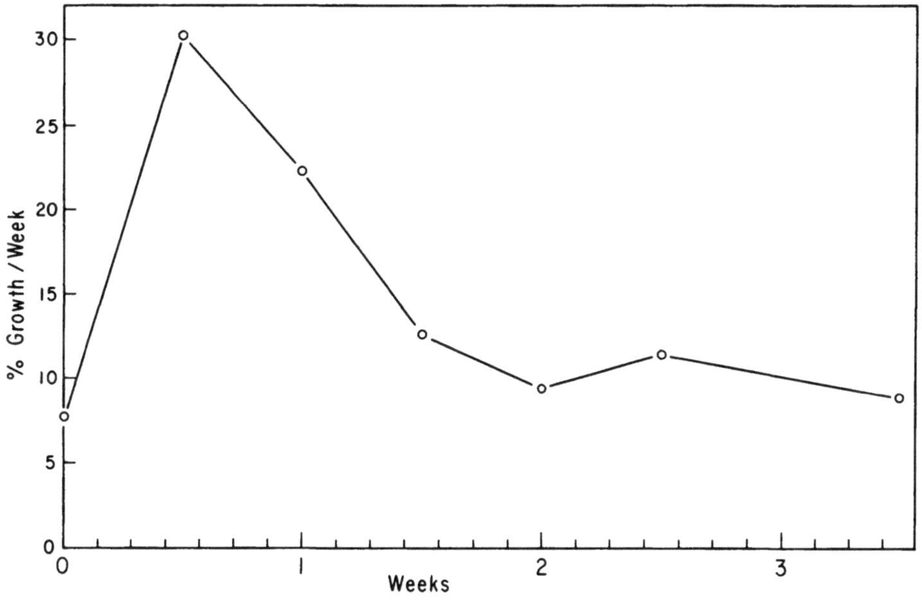

Fig. 1: Variation in growth rate of Chondrus crispus after change in growth conditions. Plants were incubated under a new set of conditions at time 0 and maintained under these conditions throughout the duration of the experiment. The previous rate of growth is indicated at time 0, and the % increase is a calculated rate based on changes in fresh weight (N.D.H. Lloyd, unpublished data).

Numerous macrophytic species have been cultured in these media, although they were originally formulated for unicellular species. At the Atlantic Research Laboratory, SWM has been employed in the culture of a wide variety of species, and for many of these the life history has been completed. Almost without exception, species that we have set out to isolate have been successfully brought into culture. Thus SWM is a satisfactory medium for general purposes, and the literature suggests equally satisfactory results with other similar media such as f/2 and ES (49). My own experience indicates that with existing media it is possible to isolate successfully into unialgal culture almost any macro-algal species. These organisms may be less fastidious than unicellular species, some of which have proved to be most recalcitrant. This conclusion, however, may have to be modified when considering axenic cultures. Experience with artificial media, such as ASP-M, has been less than satisfactory, as explained later.

Here two interjections bearing on nutritional studies are in order. Both relate to interpretation of results and pertain to experimental procedures. Firstly, stimulation of growth or other physiological responses following addition of a nutrient or a change in physical conditions may occur as a transient phenomenon such as illustrated in Fig. 1. Following, in this instance, an initial burst of growth, the response returns to the original base level although incubation continues under the new condition. Ananalogous situation is that the toxic effects of, for example, heavy metals are generally less in long-term experiments than in short-term experiments. If the response is measured soon after the addition of the metal, as is usually the case (19), a short-term experiment, extending to even a week or more, can be very misleading. This may also explain why growth data may not always be coincident with short-term photosynthetic data (6). It should be pointed out that in short-term photosynthetic experiments usually the influx of carbon, not net assimilation of carbon, is measured, and productivity data tend to be overestimates, both in culture and under field conditions.

Table VI. Relation of carbon concentration in the medium to pH and photosynthetic activity (R.G.S. Bidwell, unpublished data).

	pH	$[CO_3]$[a] mM	NPS[b] $\mu l\ CO_2\ min^{-1}\ g$ fresh wt^{-1}
Chondrus crispus			
outdoor tank, sunlight, vigorous aeration	8.3	1.5	13
as above: CO_2 added	8.0	2.0	18
Gracilaria tikvahiae			
SWM medium, 2500 lux, aeration	8.7	0.19	2.2
as above, pH adjusted with TRIS/TES	7.0	0.23	2.6
as above, pH adjusted with CO_2	8.0	1.9	18.0

[a] $[CO_3]$ in the medium measured as CO_2 released upon acidification.
[b] NPS: net photosynthesis measured as CO_2 uptake.

Secondly, and probably this is a very frequently overlooked point, there is the possibility of carbon limitation during an experiment (5). Carbon may rapidly become depleted from the medium (Table VI), both in flow-through systems and batch cultures, indeed, the medium may be deficient in carbon at the time of inoculation. This is an especially important consideration in kinetic experiments, and I suspect that at least some of the published results are equivocal because they were obtained under carbon-limiting conditions.

Individual Elements

Because of the obvious requirements for nitrogen and because the natural occurrence of this element is in low and often limiting concentrations (Table III), the major interest in nutrition of macro-algae has focused on nitrogen. Many of these investigations have been concerned with species of *Laminaria* and *Macrocystis*, as discussed elsewhere in this volume (36,77).

Nitrate-nitrogen seems to be universally assimilated by macro-algae as is nitrite (24). I am unaware of a macro-algal species unable to assimilate nitrate as are some unicellular species which require a reduced form of nitrogen for growth (30) and presumably lack nitrate reductase. However, growth and pigment development of *Pterocladia capillacea* are better with ammonium than with nitrate (24,75). Ammonium is assimilated by most species although there was no increase in length of the stalk of *Acetabularia mediterranea* with this source of nitrogen (1). At high concentrations in alkaline media, ammonium is commonly toxic (72). Nitrate may also become toxic, as I have noted for species of *Fucus* which become necrotic when in batch culture the concentration exceeds 200 µM (unpublished observations). The concentrations of both nitrogen and phosphorus, in media usually employed for unicells (49), are too high for macrophytes, and can be reduced considerably (76) as shown in Table V. In batch culture, growth of *Gracilaria tikvahiae* was saturated at 300 µM (unpublished observations), and much lower concentrations were adequate in a flow through system, for example with *Laminaria saccharina* (13, see also

77).

Nitrogen uptake by macro-algae has been considered to follow Michaelis-Menten kinetics (e.g. 20,70,71), but there are difficulties in an interpretation of some data. Michaelis-Menten kinetics were developed for enzyme-substrate reactions in vitro and their extension to macro-species may not only obscure the interpretation of uptake mechanisms by intact plants, but give the false impression that uptake of nitrogen and growth of the plant are related by a single reaction. For instance, short-term experiments, in some cases utilizing only segments or punches of thalli, have provided data which suggest that the relationship between V_{max} and K_m is fixed, whereas they are independent variables. The correct interpretation is that the demand for a nutrient is created by growth and is thus an integral of growth (14). For Michaelis-Menten kinetics, however, V_{max} and K_m must be determined under conditions of equilibrium. Conditions of nonequilibrium may in part explain the observed departure from saturation kinetics noted in some experiments (33). In addition, kinetics of multiphasic uptake have been given scant attention. Nevertheless, it is the total amount, not the concentration, of a nutrient in a system which is important, and batch cultures may utilize available nutrient resources more effectively than flow-through systems.

With some species uptake of both ammonium and nitrate is unaffected by light (70), whereas with others the rates of uptake are slower in darkness (34). Nevertheless, some assimilation seems to occur regardless of the light conditions. With many algae, especially unicellular forms, ammonium suppresses the uptake of oxidized nitrogen. The same applies to some species of macro-algae (21,33) but others can assimilate both ammonium and nitrate simultaneously (33,77). Although nitrate apparently does not suppress the uptake of ammonium, it may inhibit the uptake of nitrite (34).

The ability of macro-algae to assimilate ammonium may exceed that for other sources of nitrogen, and with Gracilaria foliifera (= G. tikvahiae) growth rates with ammonium exceeded those with nitrate (21). Indeed, the uptake of ammonium may surpass that necessary to sustain maximum growth. Certainly, simultaneous assimilation of ammonium and nitrate confers an ecological advantage, as does the luxury consumption of nitrogen (34).

Urea may serve as a sole source of nitrogen (1,24) although not all species grow well or are even able to utilize this compound, including G. tikvahiae (unpublished results). Urea is, however, difficult to assess as a source of nitrogen in bacterized culture. Other organic compounds, especially amino acids, can serve as nitrogen sources (24,57), but no single organic nitrogen compound is suitable for all species (72).

Nitrogen may affect morphology. The development of hairs is probably the most common response to low concentrations or absence of nitrogen (1,66). Consequently, high concentrations of ammonium suppressed formation of hairs in Cladosiphon zosterae (45). Hairs apparently increase the area for surface absorption, but the physiological value of this possible adaptation reamains speculative.

Embryos of species of Fucus are morphologically distinct when nitrogen deficient (Fig. 2A,D; 50). With Cladosiphon zosterae (45) the morphology of the disc is controlled by the source of nitrogen (NH_4^+ or NO_3^-). "β--Discs" are formed with nitrate or in the absence of nitrogen. A "NH_4-morphology" occurs in the presence of ammonium whether or not nitrate is also present, indicating that ammonium is preferentially assimilated. In high ammonium concentration, growth of C. zosterae occurs as callus tissue, which may serve as a possible starting point for tissue cultures.

Nitrogen deficiency suppresses reproduction of many species, but the source of nitrogen may also control reproduction, as shown for example with Acetabularia (1). More information is needed on effects of nutrients on reproduction, including nutrients other than nitrogen.

Information on phosphorus metabolism by macro-algae is scanty, although a deficiency of this nutrient is usually easily demonstrable. For example, with embryos of Fucus, development is severely retarded shortly after germination (50). Orthophosphate is the common form added to media, but most species also grow well with α- or β-glycerophosphate as the sole source of phosphorus (24). Calcification in Jania rubens was inhibited by phosphorus concentrations of 50 μg-at l^{-1} or higher and not antagonized by

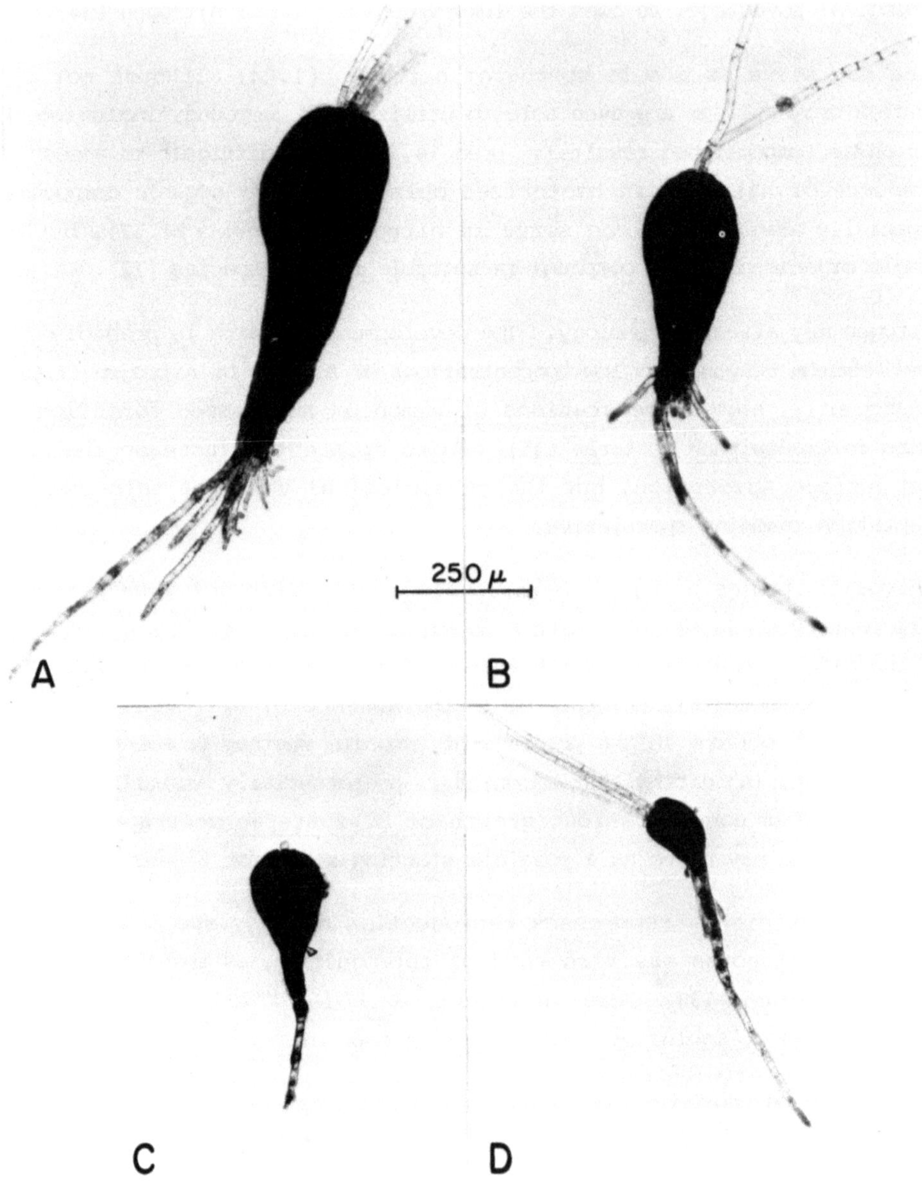

Fig. 2: Comparison of developments of embryos of Fucus edentatus in: (A) complete medium; (B) medium minus bromine; (C) medium minus boron; (D) sea water minus nitrogen (reproduced from [50] by permission of Blackwell Scientific Publ. Ltd.).

arsenite (76). Interestingly, the gametophyte was not affected by concentrations of phosphate; and, under similar conditions, inhibition of calcification was not noted in Corallina officinalis. Iron deficiency is usually demonstrable (50), although Boalch (9) failed to detect an iron requirement for Ectocarpus confervoides. The iron requirement for this species must be low and can be provided by contamination in salts, thus paralleling the situation for other heavy metals required by algae. Conceptacle formation in Jania rubens was promoted by increasing the concentration of iron or lowering the concentration of chelator (76), suggesting the possible importance of free-ion concentration in the medium.

Because of ionic interactions and salinity requirements, it is difficult to study the role of major cations in the medium. With unicellular species requirements have been demonstrated for sodium, potassium, calcium, and magnesium and the apparent required concentrations are similar to those for freshwater algae (48). Sodium is an exception; this element seems to be necessary in relatively high concentrations, and is not replaceable by any other element. But no satisfactory explanation is forthcoming for the metabolic role of sodium at these concentrations. The requirement by blue-greens and some higher plants (12) for this element is at much lower concentrations (59). Calcium is utilized in relatively high concentrations by calcified species (17), but in these instances the requirement or beneficial effects are obvious. Clearly much more work is needed on the nutritive role of these cations in macrophytes as well as unicells.

There is a similar lack of information with respect to the major anions of the medium. Sulfate can be reduced to the levels of freshwater media for growth of unicells (48). Many macro-algae contain sulfated polysaccharides as cell-wall components (61), and in these algae a major assimilation of sulfur is to be expected. For the growth of unicellular species, it has been possible to replace chloride salts of the medium with bromide salts (52). The requirement for chlorine must be as a micronutrient as in other plants, although there is little information available for macrophytes. Blinn and Markham (8) cultured gametophytes of two species of Laminariales in saline pond-water chemically similar to seawater except for about a 50-fold reduction in chloride and a 20-fold increase in sulfate. In one species the gametophytes became fertile, but no sporophyte was produced. These results should be persued, especially to determine possible effects of these ions on fertility.

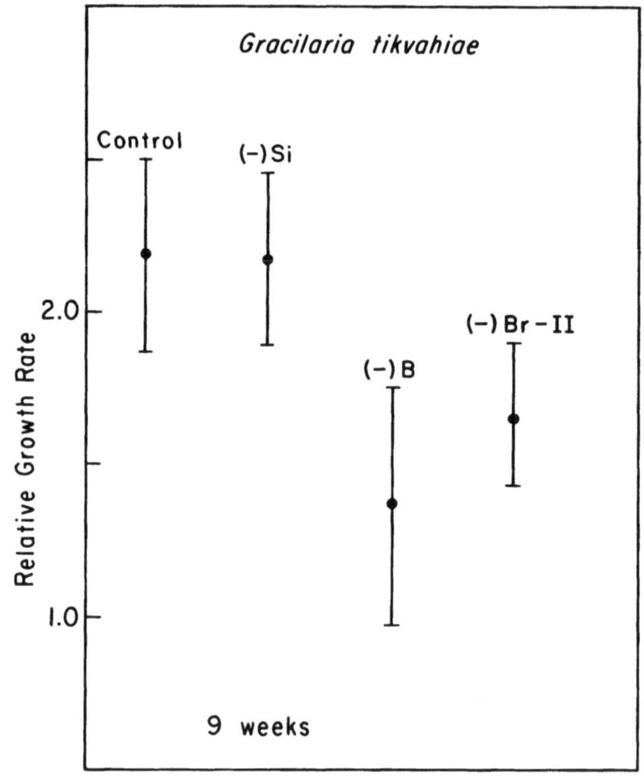

Fig. 3: Relative weekly rates of growth (fresh weight) of Gracilaria tikvahiae during 9 week in ASP-M complete and deletions as indicated. For each condition 3 replicate flasks with 40 10-mm apical segments per flask, were incubated.

Silicon is required by diatoms and other algae with siliceous structures (44). Silicon is probably universally present in the ash of algae and other plants (17,22), and this element has been recently reported to be rapidly assimilated by a number of planktonic species (27), even though it may not be required. Our own preliminary data suggest that the silicon disappearing from the medium may not necessarily be assimilated, but rather may be precipitated together with aluminum. The disappearance of soluble aluminum from seawater and culture medium has also been correlated with an increase in biomass (47,69). To reemphasize, the presence of an element in the ash of an alga does not necessarily involve assimilation, much less an implied requirement.

Silicon has not been implicated as a required nutrient for macro-algae and commonly it is omitted from the medium. With Gracilaria tikvahiae I have been singularly unsuccessful in demonstrating a requirement (Fig. 3). Similar results were obtained in culturing embryos of Fucus (50), even though germanium is toxic to these organisms (51). Other workers (54) reported a silicon requirement for Cladophora glomerata, a freshwater species, suggesting that silicon was deposited in the cell wall and was involved in protein synthesis. Their conclusions for a requirement were based largely on inhibition by germanium, and further study is necessary on silicon metabolism in this organism. Nevertheless, evidence being marshalled indicates that there may be a primary requirement for silicon by biological systems (59), albeit in minute quantities, but silicon may be a "functional" element rather than being an essential element (14). The present techniques for studying silicon will have to be improved as this element is ubiquitous, being the second most abundant element on earth, and contamination presents a serious problem.

If we accept the premise that the nutrition of algae is not unlike that of other plants, then the heavy metals listed in Table I must be regarded as essential for algae, and this has indeed been shown to be the case (59). However, none of these metals, iron excepted, has been demonstrated as essential for marine macro-algae. Uptake of heavy metals has been studied and correlated with changes in concentrations in seawater (56) as well as with stimulation of growth by fortification of the medium (58). Even so an unequivocal demonstration that an element has been assimilated is in many cases lacking, I have already mentioned the example with silicon. Thus, experiments demonstrating requirements for heavy metals by macro-algae remain to be done, and reference is made to the very recent results of Kuwabara and North (43).

About a decade ago chelating resins became available for removal of heavy metals from marine media (18), suggesting that this technique should be useful in the study of trace-metal metabolism (49). Morel et al. (55) recently used this technique with a computer program for solving simultaneous chemical equilibrium equations for describing speciation in the medium. This may be especially important because ionic species and ionic ratios, and interactions rather than individual concentrations, may be the critical factors (2,73). We should now anticipate new basic information on heavy-metal nutrition of marine macrophytes, and recent data of Kuwabara and North (43) verify this prediction. These authors investigated the effects of nine metals on the growth of gametophytes of Macrocystis pyrifera. Even in this work, the omission of several metals from the medium had no effect on growth, indicating that contamination was a problem and a common experience in many experiments.

Requirements by macrophytes for nonmetalic micronutrients (Table I) is much better documented than that for metallic elements. Although the essentiality of chlorine for algae has yet to be demonstrated, boron, iodine, and bromine must be considered essential as these elements have been shown to be required by at least one species of algae.

Iodine is the most studied of the halogens, and in some species this element may accumulate to appreciable quantities (22,83). In Laminaria japonica, for instance, iodine constitutes up to 1% of the dry weight, and the Chinese in their cultivation of L. japonica have selected strains for high iodine content (3). Although a growth stimulation of species of Laminaria by iodine was reported nearly 50 years ago by Harries (35), there is little definitive evidence that it is essential for growth. Kuwabara and North (43) indicated that iodine may be beneficial to gametophytes of M. pyrifera. An iodine requirement was reported for Asparagopsis armata (74,75) and subsequently for several other species of red (23,40) and brown (39,40,60) algae. Unquestionably, the most thorough study has been that of Woolery and Lewin (81) for Ectocarpus siliculosus. These authors demonstrated an iodine requirement of at least 1.5×10^{-3} µg-at l^{-1} for vegetative growth and of 2.5×10^{-2} µg-at l^{-1} for normal formation and maturation of plurilocular sporangia. They also demonstrated an absolute requirement for iodine by E. siliculosus by showing that iodine was not replaceable by bromine and that in the absence of iodine there was a lack of growth and apparent death of thalli. No doubt other species will be shown to have a similar requirement for iodine. Interestingly, the gametophytic phase of A. armata takes up iodine from the medium more rapidly than the tetrasporophytic (Falkenbergia) phase (42).

Bromine has also been shown to be required by several species of macro-algae. Fries (25) suggested a requirement for Polysiphonia urceolata but not for Goniotrichum alsidii. Some species of red algae

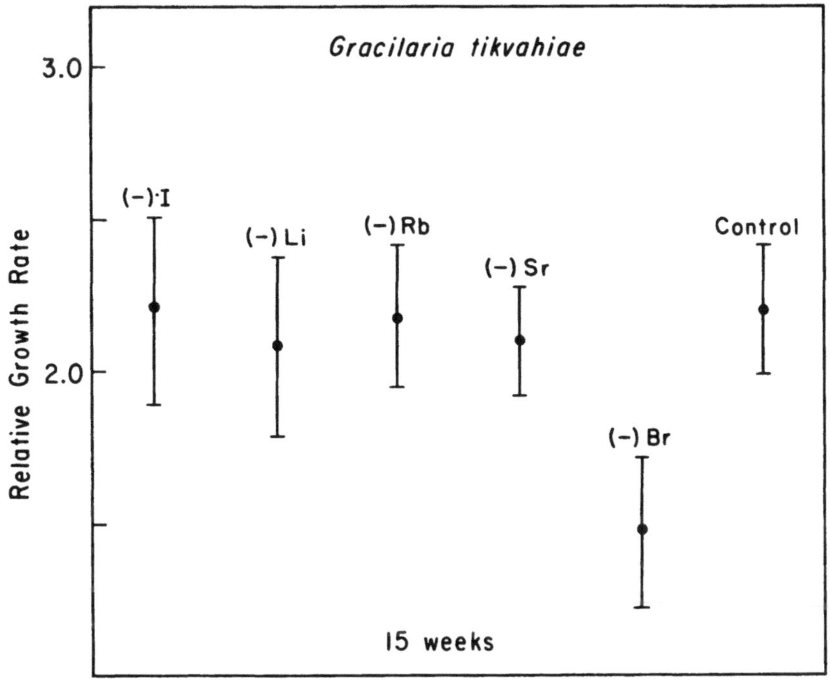

Fig. 4: Relative weekly rates of growth (fresh weight) of Gracilaria tikvahiae during 15 weeks in ASP-M complete and deletions as indicated. For each condition 3 replicate flasks with 40 10-mm apical segments per flask, were incubated.

have a relatively high content of aromatic brominated compounds, up to 6% of the dry weight (82), and 2,3-dibromo,4,5-dihydroxybenzylalcohol could replace the bromine requirement of P. urceolata (25). Growth of embryos of Fucus was considerably reduced with the omission of bromine from the medium (Fig. 2B). This condition was correctable by addition of 50 µM NaBr, even though bromine as a contamination was not eliminated from the medium (50). I have also shown a significant reduction in the rate of growth of Gracilaria tikvahiae under similar conditions (Figs. 3,4) with subsequent recovery by the addition of bromine to the medium (Fig. 5). Active uptake of bromine had previously been reported for this species (32).

Bromine, like iodine, will quite probably be shown to be required by a variety of species. Neither of these elements is known to be essential

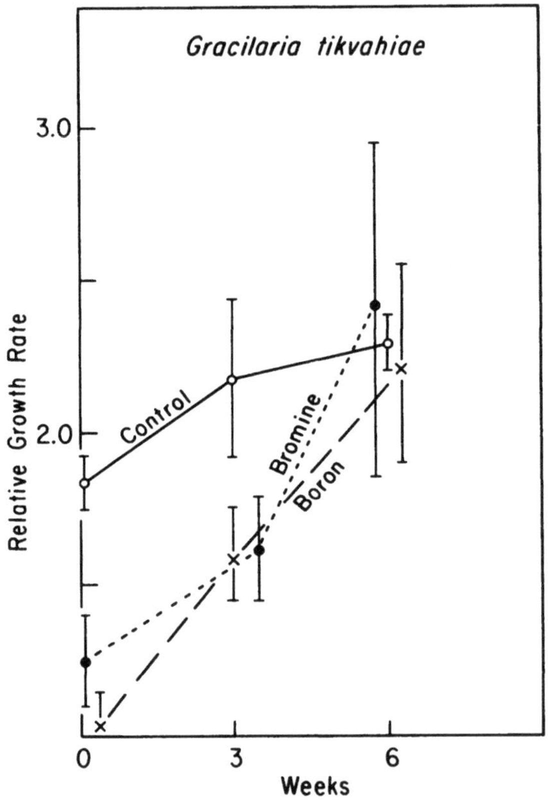

Fig. 5: Increase in rate of growth of boron- and bromine- deficient thalli of Gracilaria tikvahiae after addition of element indicated.

for freshwater species (59) or higher plants (22). Moreover, no species has been shown to require both bromine and iodine, but Codomier et al. (15) obtained a stimulation of growth with Asparagopsis armata, a species requiring iodine (74,75), at a concentration of 0.4 mM Br; at higher concentrations growth was the same as without added bromine. The conditions established by Woolery and Lewin (81) should certainly have revealed a bromine deficiency if this element were required by Ectocarpus siliculosus.

Boron is required by higher plants (14,22) and marine and freshwater diatoms (59), with evidence for a requirement by other freshwater algae being equivocal. Omission of boron from the medium resulted in moribund embryos of Fucus (Fig. 2C) and normal development of the embryos occurred with the addition of 1 µM boron (50). Growth of Gracilaria tikvahiae was significantly reduced in a "boron-free" medium, although even after several months of incubation under these conditions the plants remained viable (Fig. 3). The reduced rate of growth resulting from the omission of boron was restored by replacing boron in the medium, but with deficient plants several weeks of incubation were required to reestablish the rate (Fig. 5).

At this point it is premature to suggest that boron is required by a wide range of macro-algae. Unlike bromine, where a number of brominated organo-compounds have been identified, no analogous boronated compounds are known from algae. Similarly, with higher plants where a requirement for boron seems to be universal, the metabolic role of this element is mostly a matter of conjecture (14,22).

Artificial Media

Assuming that most, if not all, elements required by marine algae have been identified, it should be possible to formulate artificial media which are capable of indefinitely supporting growth, comparable to that in enriched media, as for higher plants. Most marine unicellular species respond well to artificial media, although some grow less well and a few grow poorly or fail completely (37,48). As suggested earlier, most marine macrophytes can also be cultured, and this should be equally possible with either enriched or artificial seawater media.

There have been only a few studies on comparison of growth of marine

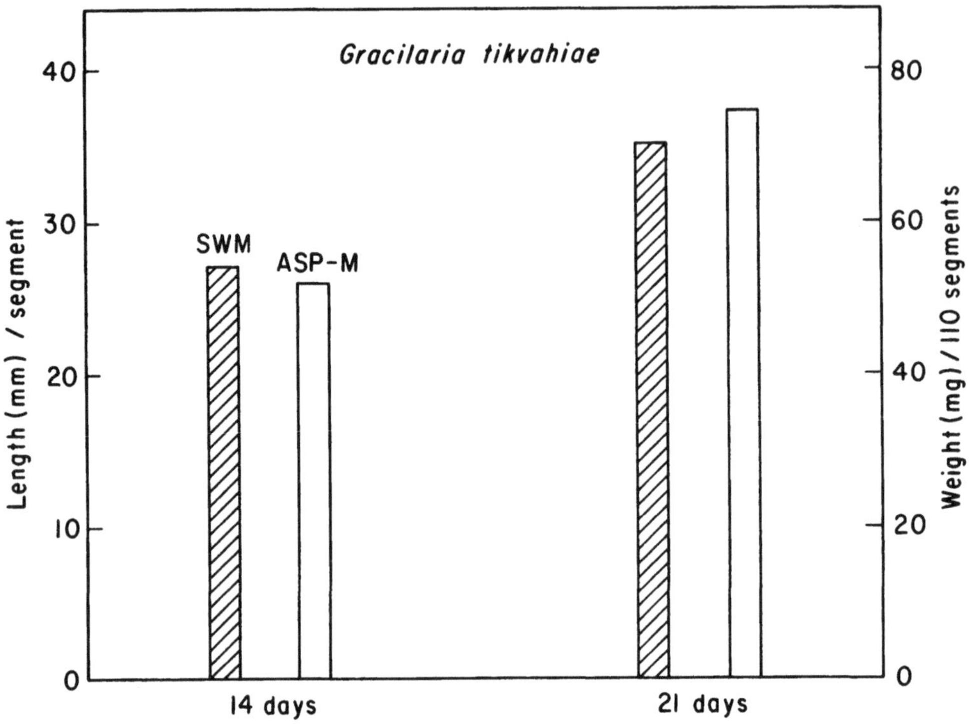

Fig. 6: Increase in length and weight of female gametophytic apices of Gracilaria tikvahiae in SWM and ASP-M media.

macrophytes in artificial and enriched seawater media. I have cultured G. tikvahiae in ASP-M for nearly two years, and based on rates of increase in length and fresh weight, and on general appearance of the plants, growth is equivalent to that in SWM (Fig. 6). Similarly, although no quantitative data are available, growth of G. foliifera also seems to be equivalent in these two media. Chondrus crispus appears to grow more slowly in ASP-M than in SWM; moreover, the plants are lighter coloured and mottled (Fig. 7). This species has, nevertheless, been in ASP-M for at least a year and large plants are available having been initiated from carpospores which following release were transferred to ASP-M. Tetrasporangia have not as yet formed, however. Furcellaria lumbricalis, like C. crispus, seems to grow more slowly in ASP-M, but the appearance of the plants remains healthy even after months of incubation. For at least three species of Fucus (F. distichus, F. edentatus, F. serratus), I have been unable to obtain continued growth in ASP-M. Reasonable growth is achievable in the first four to six weeks after transfer to ASP-M, but subsequently, necrosis begins, especially in the apical regions, and progresses with time.

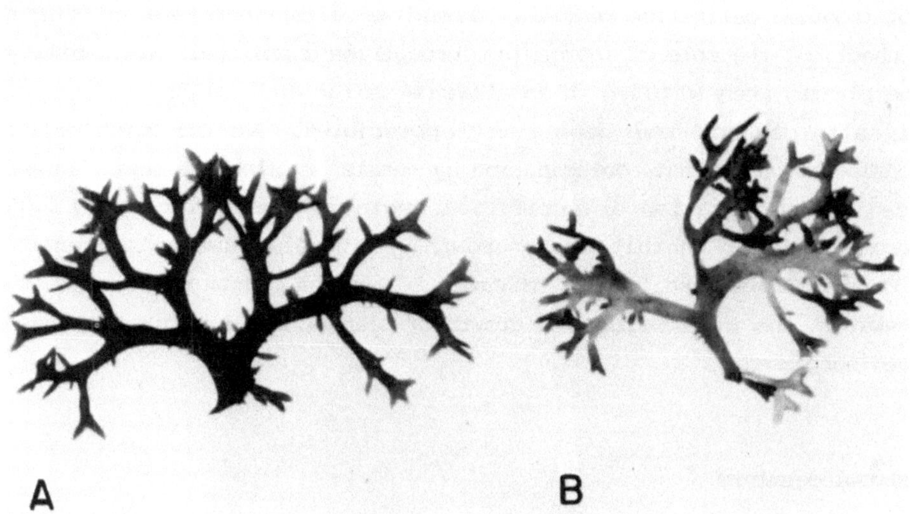

Fig. 7: Chondrus crispus incubated in: (A) SWM; (B) ASP-M media.

Similarly, embryos of Fucus can be cultured in ASP-M (50), but later become necrotic at about the time the sporelings begin to flatten. In contrast, Fries (26) seems to have successfully cultured Fucus spiralis, a species that I have not cultured, in an artificial medium without reporting necrosis. But these plants grew extremely slowly, so it is difficult to compare her results with my own. Other similar cases could be cited from the literature, and there are no apparent trends or patterns. Even for chlorophycean species, there have been problems. Keck (41) was unable to grow Acetabularia in artificial media, whereas, Shephard (67) achieved comparable growth in an enriched and an artificial medium.

In most instances, conditions of temperature, aeration, light, and so forth have been such that growth is expected to be minimal. Thus, it is impossible to assess the success of artificial media on the basis of the few data available.

Growth factors, other than vitamins listed in Table V, may be required by marine macrophytes and in part may explain the failure of artificial media. However, the addition of various organic compounds, including known plant-growth hormones, both naturally occurring and synthetic, has not produced definitive results. Should we be looking elsewhere? In this connection, the role of inorganic micronutrients and their availability to the plants, when supplied in an alkaline medium with a high concentration of salt, may not have been fully appreciated. Another point requiring consideration is that contamination by metals or other elements in salts used for preparation of artificial media may be sufficiently high to become toxic or inhibitory to growth. Natural seawater is generally "cleaner" than artificial seawater. Our goal thus remains to mimic seawater, the ideal medium for growth of plants, and a medium that has not been surpassed.

Acknowledgements

I thank R.G.S. Bidwell and N.D.H. Lloyd for permitting me to use their unpublished data.

References

1. Adamich, M., Gibor, A., Sweeney, B.M.: J. Phycol. 11, 364-367 (1975).

2. Anderson, M.A., Morel, F.M.M., Guillard, R.R.L.: Nature 276, 70-71 (1978).

3. Anonymous: Sci. Sinica 19 (1976).

4. Arnon, D.I., Stout, P.R.: Plant Physiol. 14, 371-375 (1939).

5. Bidwell, R.G.S.: In Aspects of nitrogen and carbon metabolism; A symposium on the physiology and Biochemistry of Plant Productivity, J.D. Dewley, ed.), W. Junk, The Hague, Netherlands, pp 1-17, 1981

6. Bird, N.L., Chen, L.C.M., McLachlan, J.: Bot. Mar. 22, 521-527 (1979).

7. Black, W.A.P., Mitchell, R.L.: J. mar. Biol. Ass. UK 30, 575-584 (1951).

8. Blinn, D.W., Markham, J.W.: Phycologia 8, 51-55 (1969).

9. Boalch, G.T.: J. mar. Biol. Ass. UK 41, 287-304 (1961).

10. Bold, H.C.: Ann. Missouri Bot. Gard. 61, 14-44 (1974).

11. Brewer, P.G.: In Chemical Oceanography (2nd ed.), Vol. 1, (J.P. Riley, G. Skirrow, eds.), Academic Press, London, pp. 415-496, 1975

12. Brownell, P.F.: In Advances in Botanical Research, Vol. 7 (H.W. Woodhouse, ed.), Academic Press, New York, pp. 117-224, 1979

13. Chapman, A.R.O., Markham, J.W., Lüning, K.: J. Phycol. 14, 195-198 (1978).

14. Clarkson, D.T., Hanson, J.B.: Ann. Rev. Plant Physiol. 31, 239-298 (1980).

15. Codomier, L., Segot, M., Terte, J., Jeanty, G.: Soc. Phycol. France Bull. 23, 29-37 (1978).

16. Collièr, A.W.: In Marine Ecology, Vol. 1, pt. 1, (O. Kinne, ed.), Wiley-Interscience, London, pp. 1-93, 1970

17. Darley, W.M.: In Algal Physiology and Biochemistry, (W.D.P. Stewart, ed.), Blackwell, Oxford, pp. 655-675, 1974

18. Davey, E.W., Gentile, J.W., Erickson, S.J., Betzer, P.: Limnol. Oceanogr. 18, 993-997 (1970).

19. Davies, A.G.: Adv. Mar. Biol. 15, 381-508 (1978).

20. DeBoer, J.A., Guigli, H.J., Israel, T.L., D'Elia, C.F.: J. Phycol. 14, 261-266 (1978).

21. D'Elia, C.F., DeBoer, J.A.: J. Phycol. 14, 266-272 (1978).

22. Epstein, E.: Mineral Nutrition of Plants: Principles and Perspectives. John Wiley & Sons, New York, 1972

23. Fries, L.: Physiol. Plant. 19, 800-808 (1966).

24. Fries, L.: Bot. Mar. 16, 19-31 (1973).

25. Fries, L.: Z. Pflanzenphysiol. 76, 366-368 (1975).

26. Fries, L.: Phycologia 16, 451-455 (1977).

27. Fuhrman, J.A., Chisholm, S.W., Guillard, R.R.L.: Nature 272, 244-246 (1978).

28. Galum, E., Torrey, J.G.: Develop. Biol. 19, 447-459 (1969).

29. Goldberg, E.D.: In The Sea, Vol. II, (M.N. Hill, ed.), Interscience, New York, pp. 3-25, 1963

30. Guillard, R.R.L.: In Culture of Marine Invertebrate Animals (W.L. Smith, M.H. Chanley, eds.), Plenum Publishing Co., New York, pp. 29-60, 1975

31. Guillard, R.R.L., Ryther, J.H.: Can. J. Microbiol. 8, 229-239 (1962).

32. Gutnecht, J.: Biol. Bull. 129, 495-509 (1965).

33. Haines, K.C., Wheeler, P.A.: J. Phycol. 14, 319-324 (1978).

34. Hanisak, M.D., Harlin, M.M.: J. Phycol. 14, 450-454 (1978).

35. Harries, R.: Ann. Bot. 46, 893-928 (1932).

36. Harrison, P.J., Druehl, L.D.: In Synthetic and Degradative Aspects of Marine Macrophytes (L.M. Srivastava, ed.), Walter De Gruyter, Berlin 1981

37. Harrison, P.J., Waters, R.E., Taylor, F.J.R.: J. Phycol. 16, 28-35 (1980).

38. Hsiao, S.I.C.: Can. J. Bot. 47, 1611-1616 (1969).

39. Hsiao, S.I.C.: Can. J. Bot. 48, 1359-1361 (1970).

40. Iwasaki, H.: J. Phycol. 3, 30-34 (1967).

41. Keck, K.: In Methods in Cell Physiology, Vol. 1 (D.M. Prescott, ed.), Academic Press, New York, pp. 189-213, 1964

42. Knappe, J.: Ber. Deutsch Bot. Ges. 92, 473-482 (1979).

43. Kuwabara, J.S., North, W.J.: J. Phycol. (in press) (1981).

44. Lewin, J.C., Reimann, B.E.F.: Ann. Rev. Plant Physiol. 20, 289-304 (1969).

45. Lockhart, J.C.: Am. J. Bot. 66, 836-844 (1979).

46. Lunde, G.: J. Sci. Fd. Agric. 21, 416-418 (1970).

47. MacKenzie, F.T., Stoffyn, M., Wollast, R.: Science 199, 630-682 (1978).

48. McLachlan, J.: Can. J. Microbiol. 10, 769-782 (1964).

49. McLachlan, J.: In Handbook of Phycological Methods, Vol. I (J.R. Stein, ed.), Cambridge University Press, Cambridge, pp. 25-51, 1973

50. McLachlan, J.: Phycologia 16, 329-338 (1977).

51. McLachlan, J., Chen, L.C.M., Edelstein, T.: Can. J. Bot. 49, 1463-1469 (1971).

52. McLachlan, J., Craigie, J.S.: Nature 214, 604-605 (1967).

53. Mann, K.H., Chapman, A.R.O., Gagné, J.A.: In Synthetic and Degradative Aspects of Marine Macrophytes (L.M. Srivastava, ed.), Walter De Gruyter, Berlin 1981

54. Moore, L.F., Traquair, J.A.: Planta 128, 179-182 (1976).

55. Morel, F.M.M., Rueter, J.G., Anderson, D.M., Guillard, R.R.L.: J. Phycol. 15, 135-141 (1979).

56. Myklestad, S., Eide, I., Melsom, S.: Proc. Int. Seaweed Symp. 9, 143-151 (1979).

57. Neilson, A.H., Larsson, T.: Physiol. Plant. 48, 542-543 (1980).

58. North, W.J., Wheeler, P.A.: Proc. Int. Seaweed Symp. 9, 67-78 (1979).

59. O'Kelley, J.C.: In Algal Physiology and Biochemistry (W.D.P. Stewart, ed.), Cambridge University Press, Cambridge, pp. 610-635, 1974

60. Pedersén, M.: Physiol. Plant. 22, 680-685 (1969).

61. Percival, E., McDowell, R.H.: Chemistry and Enzymology of Marine Algal Polysaccharides, Academic Press, London 1967

62. Provasoli, L.: Proc. Int. Seaweed Symp. 4, 9-17 (1964).

63. Provasoli, L.: In Culture and Collections of Algae (A. Watanabe, A. Hattori, eds.), Japanese Soc. Plant Physiol., pp. 63-75 1968

64. Provasoli, L., McLaughlin, J.J.A., Droop, M.R.: Arch. Mikrobiol. 25, 392-428 (1957).

65. Schreiber, E.: Wiss. Meeresuntersuch. N.F. 16, 1-34 (1927).

66. Schonbeck, M.W., Norton, T.A.: Bot. Mar. 22, 233-236 (1979).

67. Shephard, D.C.: In Methods in Cell Physiology, Vol. 4 (D.M. Prescott, ed.), Academic Press, New York, pp. 49-69 1970

68. Starr, R.C.: Proc. Int. Seaweed Symp. 7, 3-6 (1972).

69. Stoffyn, M.: Science 203, 651-653 (1979).

70. Topinka, J.A.: J. Phycol. 14, 241-247 (1978).

71. Topinka, J.A., Robbins, J.V.: Limnol. Oceanogr. 21, 659-664 (1976).

72. Ukeles, R.: In Marine Ecology Vol. III, pt. 1 (O. Kinne, ed.), John Wiley and Sons, London, pp.367-466 1976

73. Underwood, E.J.: Prog. Water Tech. 11, 33-45 (1979).

74. von Stosch, H.A.: Naturwissenschaften 49, 42-43 (1962).

75. von Stosch, H.A.: Proc. Int. Seaweed Symp. 4, 142-150 (1964).

76. von Stosch, H.A.: Proc. Int. Seaweed Symp. 6, 389-399 (1969).

77. Wheeler, W.N.: In Synthetic and Degradative Aspects of Marine Macrophytes (L.M. Srivastava, ed.), Walter De Gruyter, Berlin 1981

78. Whyte, J.N.C., Englar, R.J.: Tech. Rept. Fish. Mar. Ser. Ottawa 509, (1974).

79. Whyte, J.N.C., Englar, J.R.: Bot. Mar. 23, 13-17 (1980).

80. Whyte, J.N.C., Englar, J.R.: Bot. Mar. 23, 19-24 (1980).

81. Woolery, M.L., Lewin, R.A.: Phycologia 12, 131-138 (1973).

82. Young, D.N., Howard, B.M., Fenical, W.: J. Phycol. 16, 182-185 (1980).

83. Young, E.G., Langille, W.M.: Can. J. Bot. 36, 301-310 (1958).

DISCUSSION

BIDWELL: I have a comment related to Dr. Willenbrink's observations this morning about C_4 fixation in photosynthesis of marine plants. The observation of sodium as a required nutrient is interesting because in higher plants sodium is a required nutrient only for those plants which have a C_4 photosynthetic system, such as conventional C_4 plants and crassulacean plants. These plants always have PePC and usually the PePCk and the inference is that the sodium requirement is related to the C_4 - carboxylation. The interesting thing is that sodium is a required element for the blue-green algae also, but I do not think that they are C_4 plants; they do have an active C_4 metabolism, however.

NORTH: I would like to emphasize the importance of contamination in trace metal studies. One of our graduate students made a calculation that the amount of zinc in submicron particles contained in 1 liter of air in an ordinary urban environment, if allowed to settle in a petri dish with normal sea water, could double the amount of zinc in that water.

McLACHLAN: It is not necessarily the concentration of the elements , but the form in which they exist relative to each other that is important. It is this interaction that we have probably not appreciated sufficiently in the past. Now there are computer programs available and these calculations can be made. This will help a lot.

FOSTER: Another problem seems to be in the heavy metal contamination of the reagents that are used. In some work done in our laboratory we find routinely that any culture we get from the culture collection, the heavy metal content and everything else is on the order of one magnitude higher than the heavy metal content in the ocean.

McLACHLAN: I think it would now be possible (with the use of computers, Ed.) to correct for these errors. Also there have been available for some time chelated resins and I think they can be used to clean up some of these media with very high salt concentrations, something we have not been able to do before.

NUTRIENT UPTAKE AND GROWTH IN THE LAMINARIALES AND OTHER MACROPHYTES: A CONSIDERATION OF METHODS

P.J. Harrison

Departments of Botany and Oceanography, University of British Columbia, Vancouver, B.C. V6T 2B1, Canada

L.D. Druehl
Bamfield Marine Station, Bamfield, B.C. V0R 1B0, Canada

Introduction

The growth rate of seaweeds may be limited by several environmental factors such as light, temperature and nutrients. The latter factor has just begun to receive attention and this review will focus only on nutrient limited growth of marine macrophytes, particularly kelp.

Studies of the nutrient requirements of seaweeds indicate that while over 50 elements may be found in seaweeds (8), less than half of these may actually be required for growth. Most of those elements which are required for growth are present in seawater in excess amounts; however, a few elements such as N, P, or Fe may limit growth at certain times because they are present in low concentrations or, as in the case of iron, are in a biologically unavailable form. These three elements may limit the yield (biomass) as well as the growth rate (metabolic reactions) of the seaweed. This latter aspect has been studied quantitatively by examining one component of it, the relationship between the ambient concentration and rate of uptake of the limiting nutrient.

Nutrients generally enter plant cells as either cations or anions. These ions may move across the cell membrane by passive diffusion (i.e. down an electrochemical gradient), or by active transport, a process requiring energy since movement is against an electrochemical gradient. The latter uptake mechanism is thought to be the most important for the limiting nutrients because the external or ambient concentration is generally

Synthetic and Degradative Processes in Marine Macrophytes

several orders of magnitude less than the intracellular concentration. This active uptake mechanism exhibits saturation kinetics; at low nutrient concentrations the uptake rate increases linearly with increasing concentration, however, at high nutrient concentrations a further increase in nutrient concentration does not result in any further increase in the uptake rate. The relationship between uptake rate of the limiting nutrient and its concentration is generally described by a rectangular hyperbola, similar to the Michaelis-Menten equation for enzyme kinetics (Fig. 1A). The equation is:

$$V = V_{max} \frac{S}{K_s + S}$$

where V and V_{max} are the uptake and maximal uptake rates, respectively, and S is the substrate (limiting nutrient) concentration. K_s is called the half-saturation constant and it is the substrate concentration where $V = V_{max}/2$. The lower the value of K_s for a particular macrophyte, the higher is the affinity of the carrier sites for the nutrient. The two uptake kinetic parameters in the above equation which are routinely used to describe the uptake characteristics of a particular plant, are K_s and V_{max}. Recently, it has been suggested that the slope of the initial part of the hyperbola may be a more useful parameter for comparing competitive ability for the limiting nutrient, rather than the K_s value, because in the latter case, when V_{max} increases or decreases the value of the K_s automatically changes also (24). The kinetic parameters may be calculated more accurately by rearranging the Michaelis-Menten equation above to yield a straight line from which K_s and V_{max} can be calculated from the slope or axis intercept. The advantages and disadvantages of the three possible linear plots have been examined by Dowd and Riggs (11) and the $^S/_V$ _vs._ S or the $^V/_S$ _vs._ V plots are generally superior to the $^1/_S$ _vs._ $^1/_V$ plots.

Active uptake may not follow the simple saturation kinetics described above. Studies with higher plants have revealed that the pattern of uptake may be biphasic or multiphasic (13,32). To date there is no good evidence for such a pattern in macrophytes. Macrophytes may take up the

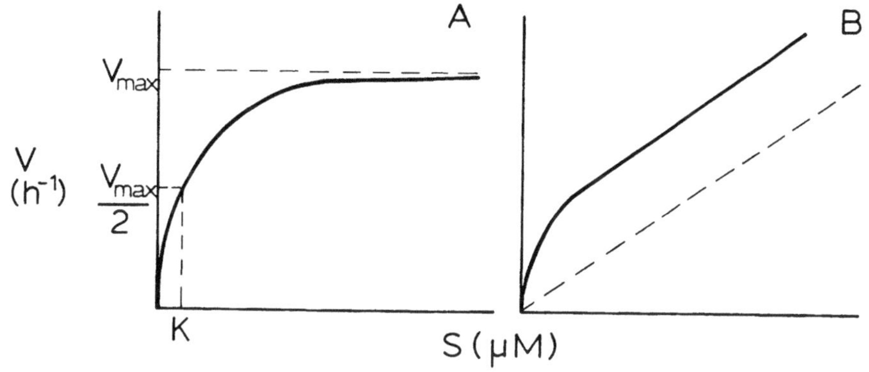

Fig. 1: Hypothetical plots of nutrient uptake rate (V) and concentration of the limiting nutrient (S), showing: A) active uptake only, where V_{max} is the maximal uptake rate and K_s, the half-saturation constant, which is the concentration where $V = V_{max}/2$ and, B) active uptake plus a linear diffusive component (solid line) and only a diffusive component (dotted line).

limiting nutrient by a secondary process, diffusion. Its relationship with the concentration of the limiting nutrient is shown in Fig. 1B. In the case of simple diffusion, the rate of uptake is generally proportional to its external concentration (i.e. a linear relationship even at high nutrient concentrations). Diffusion across the cell membrane is not likely to be important, except at concentrations greatly exceeeding those found environmentally (9,17). It may be important for the movement of the nutrient ion into the apparent free space or apoplast (13,29) which includes the cell walls and the intercellular spaces (16,18). The volume of the apparent free space may be considerable, with values of up to 20 to 30% of the cell volume for the red alga, Porphyra perforata (12). When an alga is placed in a nutrient medium an initial rapid uptake of ions may occur for a few minutes and this uptake is considered to occur by diffusion into the apparent free space.

Of all the nutrients required for growth of macrophytes, nitrogen is regarded as the most commonly limiting nutrient of coastal surface waters during the summer months (36). The ability to procure or compete for the limiting nutrient or to mobilize intracellular reserves during this nitrogen limited summer period, may partially determine the success of a macrophyte in the benthic community. For this reason, there has been considerable interest in determining the uptake kinetic parameters, V_{max} and K_s, for nitrogen in order to quantitatively assess the ability of an alga to compete for this limiting resource. For example, a good competitor would have a high V_{max} and a low K_s value. The number of studies on nitrogen uptake by macrophytes greatly lags those for phytoplankton.

Only a few phaeophytes (5,17,21,44,50) and other macrophytes (9,19,20) have been studied, but already there is considerable variability in the methods of assessing uptake rates.

Uptake *vs.* Growth

The specific uptake rate (V) is equal to the specific growth rate (u) when nutrients are not limiting, or during steady state growth under nutrient limiting conditions. From numerous experiments with marine phytoplankton (2,6), it has been shown that under transient conditions when the limiting nutrient is provided to an alga growing under nutrient limitation, $V > u$ until the amount of the limiting nutrient per cell, or the cell quota, Q, reaches its normal (maximal) concentration. Therefore, during periods of nutrient limitation, the *transient* maximal uptake rate, V_{max}, which is measured during short term nutrient uptake experiments, would over-estimate the maximal growth rate, u_{max}, by several times (9). The half-saturation constant, K_s, determined during transient conditions of short term uptake experiments, may also be greater than that for growth. For two red algae, DeBoer *et al.* (7) found the K_s for ammonium limited growth was very low, ranging from 0.2 to 0.4 µM. Similarly, for some kelp such as *Laminaria saccharina* the K_s for nitrate limited growth was 1.3 to 1.5 µM (5) while the K_s for uptake ranged from 4 to 6 µM nitrate for *Laminaria longicruris* (21).

Uptake rate is a good approximation of growth rate when nutrients do not limit growth, or when steady state growth occurs under nutrient limiting conditions. Under nutrient limited conditions uptake rate is an assessment of an alga's short term response under transient conditions, while the growth rate represents a long term response and consequently it is more of an integrating parameter. Both measurements are really necessary to understand an alga's total response *in situ*; in the past there has been a bias towards measuring only nutrient uptake rates.

Growth rate measurements will not be covered in detail as these have recently been reviewed by Chapman (3). Briefly these methods involve destructive sampling for changes in dry weight, and plant carbon or nitrogen, or nondestructive sampling which include changes in wet weight, surface area or the movement of punched holes away from the meristem in the case of some kelp. For ecological or field purposes, non-destructive sampling is preferable since the growth of the same plant can be followed over time and related to the limiting nutrient concentration of the ambient water. Ideally, one would like to relate growth rate to another rate parameter, the specific flux of the limiting nutrient, rather than a standing stock parameter, nutrient concentration. This relationship can only be accurately determined in continuous culture in the laboratory and the thorough study by DeBoer *et al.* (7) is a good example of how this is achieved.

The use of different methods of measuring uptake rates makes comparison of data difficult or in some cases impossible. Therefore at this early phase of studies with marine macrophytes it would be desirable to adopt a more standard procedure for determining the nutrient uptake rate. This review will consider various aspects of the methods for determining nutrient uptake and make recommendations wherever possible.

Considerations

While the reasons for making nutrient uptake measurements may vary, for the purposes of this review, the methods which are discussed will be primarily for an ecological measurement of uptake rate, rather than solely for the elucidation of physiological mechanisms. The latter has been thoroughly reviewed elsewhere (14,29,48).

I. Experimental considerations

1. Normalization of uptake rates. The four commonly used normalizations are: 1) area (μmol cm^{-2} h^{-1}), 2) wet weight (μmol g wet wt^{-1} h^{-1}), 3) dry weight (μmol g dry wt^{-1} h^{-1}), 4) specific uptake (μmol $μmol^{-1}$ particulate nutrient in the plant h^{-1}, which simplifies to h^{-1}). Since normalization by area, weight and composition of particulate matter are not interconvertible, it is advisable to report in more than one unit or at least give conversion factors, so that data in the literature may be compared. Normalization on an area basis is a useful description of the uptake process since the use of these units makes some attempt to relate the rate at which ions are taken up across a unit surface area of cell membrane. Dry or wet weights are used to normalize uptake rate on a biomass basis. The specific uptake rate is useful in allowing comparison of rates between diverse groups of organisms and estimating the effects of pre-history (i.e. response of a plant from a nutrient depleted environment vs. one from a nutrient rich environment). Since surface area and wet and dry weights are constant throughout an experiment, a different pattern of uptake vs. time or substrate will be obtained than when uptake is divided by the amount of particulate nutrient in the plant, which increases during the uptake experiment as the alga takes up the limiting nutrient. Although in some cases a constant amount of particulate nutrient in the plant may be used in the calculation of specific uptake rate if the incubation period is short, this is usually the amount at the beginning of the experiment. At present it is unclear just how quickly the uptake rate of the limiting nutrient is affected by the limiting nutrient taken up during the experiment.

2. Field vs. laboratory measurements. Ideally one would like to make the uptake measurements for autecological studies in the field, since one does not have to contend with the difficulties of simulating important physical (light quality and quantity, temperature, and water movement) and chemical (nutrient concentration and pH) factors which are characteristic of the field. To simulate field conditions, several studies have used an in situ approach for carbon uptake or oxygen evolution measurements of primary productivity by enclosing the macrophyte in a plastic bag (27,45), or large plastic cylinder (23). This approach is most suitable for

isotopes (e.g. ^{15}N, ^{32}P, ^{14}C) because small amounts of uptake can be detected. One of the problems in using isotopes is that different parts of the thallus accumulate the isotope at different rates and samples from different parts of the thallus must be taken and averaged in order to obtain a whole-thallus uptake measurement (the most useful measurement for ecological purposes). Nutrient uptake measurements can also be determined by following the disappearance of the limiting nutrient from the medium, provided the ratio of plant biomass to medium volume is high enough to produce a relatively high rate of nutrient removal.

The major disadvantages of field experiments are: 1) varying natural conditions from day to day (e.g. insolation), and 2) the difficulty in producing a significant number of simultaneous replicates. These problems are overcome under laboratory conditions; however, extrapolation to the natural environment is often difficult.

3. Laboratory conditions. If uptake experiments are conducted in the laboratory, then every effort should be made to simulate the environmental conditions in the field. Since it is difficult to reproduce the high irradiance values obtained outdoors, it is preferable to conduct a nutrient uptake or photosynthetic rate *vs.* irradiance experiment to determine at what light levels these metabolic processes are saturated. Saturation generally varies between 30 and 100 $\mu E\ m^{-2}\ s^{-1}$, however, an irradiance above 100 $\mu E\ m^{-2}\ s^{-1}$ should be saturating for most species. There have been no experiments to date testing the effect of light quality on nutrient uptake rates. If fluorescent lights are used, daylight fluorescent tubes are preferable to the more frequently used cool white tubes, because their spectrum simulates underwater light more closely. The spectral distribution can be corrected further to approximate the spectral distribution of 5 m underwater light (Jerlov Type 3), by inserting a sheet of 1/8" thick blue Plexiglas (Rohm and Haas #2069) between the lights and the plants (22).

The photoperiod should also simulate natural conditions. This is important for longer term experiments where diel periodicity in uptake may influence results. If the C/N ratio (atomic) of the macrophyte is <10, indicating nitrogen saturating conditions, diel periodicity in both nitrate and ammonium uptake may occur (9). Under nitrogen limiting conditions when the C/N ratio is >10, the uptake rate of nitrate or ammonium during the day and night may be equal (i.e. no periodicity).

Differences in water temperature between the laboratory and the field may generally be corrected by assuming a Q_{10} of approximately 2 for temperatures less than about 15 C (28). This may not be true for short term transient changes in temperature which may occur when short term uptake experiments are conducted at a different temperature than those encountered in the field.

4. Medium. Filtered natural seawater (0.45 µm filter to remove phytoplankton), preferably from the collection site, should be used. If the nutrient under study is present in very low concentrations, then the filtered medium may be enriched with known amounts of the limiting nutrient for the uptake experiment. When the filtered medium contains saturating amounts of the nutrient of interest, this nutrient may be "stripped out" by putting a large biomass of the macrophyte into the medium for several hours; these plants should not be used for the uptake experiments. It should be noted that many Phaeophyceae may be unsuitable since they often release substances which discolor the water during the nutrient stripping process. All other nutrients, vitamins and trace metals must remain saturating throughout the uptake experiment. These can be added by making up an enrichment solution such as ES (35) minus the nutrient of interest. For nutrients such as nitrogen one must choose which form (e.g. nitrate, nitrite, ammonium, or some organic nitrogen compound) to add. In the field under summer conditions when nitrogen may be limiting, uptake of ammonium, produced through regeneration primarily by herbivores, may be more important ecologically than nitrate. If uptake rates of ammonium are studied, rates for ammonium concentrations >10 µM are not of ecological interest; in fact, concentrations over 30 µM have been reported to be toxic for some species of macrophytes (47). One must also bear in mind that uptake rates of ammonium may be much greater (up to 50%) than nitrate uptake rates (9) and therefore caution must be exercised in extrapolating these high uptake rates to other seasons of the year when nitrate may be the principle form taken up.

Uptake studies conducted in the fall, winter or early spring months should use nitrate as the most ecologically appropriate form of nitrogen; however, in the summer, pulses of nitrate may also be added to the euphotic zone during periods of storm activity or tidal mixing (42). Concentrations of up to 30 µM are of ecological interest.

Organic nitrogen sources have seldom been examined. Urea has been most frequently studied and inferior growth was reported in most instances (7). The uptake of 17 different amino acids by the brown alga, Giffordia mitchellae was tested (37). Nine amino acids were taken up and leucine had the highest rate. Calculations revealed that at natural amino acid concentrations (ca. 1 µM), uptake of amino acids, at least under the conditions used in the study, would only supply a few percent of that alga's nitrogen requirements. Most studies to date have presented the macrophyte with only one organic nitrogen source at a time. This does not simulate natural conditions. Further studies are required on mixed substrates (i.e. inorganic plus organic, or several organic compounds added together). Nutrient preconditioning of the alga is also very important in determining substrate preferences. For example, in experiments with phytoplankton, uptake of amino acids was enhanced if the phytoplankton had been previously growing under nitrogen limiting conditions (51).

If uptake rates are determined by disappearance of nutrients, then enough biomass must be used to be able to measure a decrease in the nutrient concentration in the medium in a reasonable length of time. However, too much biomass may decrease the uptake rate due to decreased light absorption as a result of overlapping thalli, or the rapid exhaustion of the limiting nutrient before the end of the experiment. The above criteria can be accommodated with a biomass of approximately 5 g wet wt l^{-1} for most species, however, much higher concentrations may be used depending on the thickness and form of the thallus (e.g. filamentous vs. blade). Johnston (25) recommends a dry seaweed/water volume ratio of 0.1 to 0.3 for 24 h incubations to ensure no nutrient or CO_2 deficiencies.

5. Collection. Collection of plants should be carried out carefully in order to incur a minimum of stress, particularly light and temperature

shocks. It is preferable to collect whole plants detached at the holdfast. Storage in water in a covered ice chest and immediate return to the laboratory is desired. If experiments are conducted over a series of days, collections should be made at the same time every day in order to minimize any diel periodicity effects. Every attempt should be made to collect epiphyte-free thalli. If this is not possible then epiphytes may be gently removed by wiping the thallus. A compound microscope may be useful to determine the completeness of removal. The effect of the epiphyte will vary with the species of epiphyte, and the type and life history of the host. For example, the pennate diatom, Navicula incerta adversely influences the growth and even the survival of germlings of Fucus spiralis (39).

Bacterial contamination is more difficult to control. Antibiotics such as streptomycin and penicillin have been employed to inhibit bacterial uptake (21), but the concentration where there is no effect on algal uptake may vary according to the age or nutrient past history of the alga. It is unlikely that there is an antibiotic concentration which completely inhibits the bacteria but has no effect on the alga.

6. Experimental design. There are two basic approaches to determine the uptake kinetics of an alga. The first and the most common is to monitor the depletion of the nutrient from the alga's incubation medium. This approach gives an estimate of the net uptake (gross uptake minus excretion). However, this relies on the assumption that the excreted nutrient is in the same form as the monitored nutrient being taken up. The second approach, which to our knowledge has seldom been employed for seaweeds, is to monitor the isotopic nutrient accumulated by the plant. The disadvantage of this method relates to the assumption that the isotope/nonisotope excreted nutrient will have the same ratio (e.g. $^{15}N/^{14}N$) as the gross uptake ratio. This same problem plagues researchers using ^{14}C for productivity estimates. In this review we will focus on monitoring the depletion of nonisotopic nutrients from the medium to estimate net uptake since it is the simplest method for ecological studies. However, studies using isotopes are required to enhance our knowledge of nutrient excretion, residency time of the nutrient within the plant tissue and elucidation of physiological mechanisms.

Using the method of uptake of nonisotopic nutrient, there are two basic experimental designs in use. The simplest arrangement is to place one or several plants in high concentration of the limiting nutrient (e.g. 30 µM NO_3^-) and follow the decrease in the limiting nutrient with time (Fig. 2A). This method has been frequently used to measure uptake rates for phytoplankton and is referred to as the perturbation method (2,6). It consists of removing small aliquots of medium on a time series basis until the nutrient has been exhausted or no more is taken up. Nutrient samples can be analyzed by hand (41), although an Autoanalyzer has the advantage of a smaller sample size and nearly real time read-out of results. It may take many hours for the limiting nutrient to be taken up, and if the plants have been growing under a day/night cycle, interpretation of the pattern of uptake can be complicated by a diel periodicity in the uptake rate. To overcome this, it is recommended that a higher biomass be used to reduce the time necessary to complete the experiment. However, one must assure that carbon does not become limiting under the higher biomass conditions.

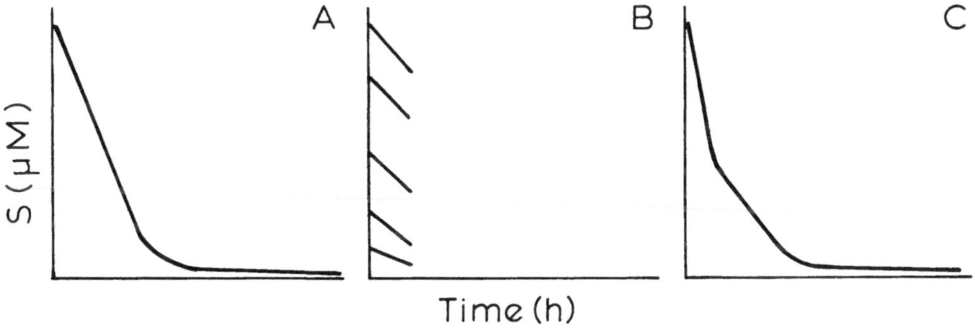

Fig. 2: Hypothetical time series of nutrient disappearance from the medium using: A) the perturbation method where saturated uptake is linear with time, B) the multiple container - constant incubation time method, and C) the perturbation method where saturated uptake is nonlinear with time.

The second method of determining uptake rates involves the use of many containers filled with filtered medium, enriched with the limiting nutrient in a series of concentrations (e.g. 2, 4, 6, 8, 12, 16, 20, and 30 µM NO_3^-). A different plant is placed in each container and the disappearance of the limiting nutrient is determined after a defined period of time (Fig. 2B). Generally, the incubation time is short (e.g. 10 to 60 min) and constant for each concentration. An alternative to this method is to use a long incubation time for the higher concentrations and a shorter time for the lower concentrations of the limiting nutrient. The variable incubation time has the serious disadvantage of overemphasizing the uptake rate at low nutrient concentrations if the uptake rate is not linear with time (Fig. 2C). This nonlinearity of uptake with time has been observed for some phytoplankton (6) and in some higher plants (26,29). The variable incubation time method does not offer any advantages in assessing uptake rate and it is not recommended for use under any circumstances.

In the Introduction of this review the possibility of uptake occurring by passive diffusion, as opposed to active uptake, was discussed. There are several ways to eliminate or estimate the contribution of diffusion of the nutrient into the apparent free space or apoplast. This initial component of uptake can be separated from active transport of the nutrient across the plasmalemma by estimating the time required for the nutrient to diffuse into and saturate the apparent free space. This can be accomplished by using a radioactive tracer such as ^{45}Ca. The plant tissue is placed in radioactively labelled medium for several hours and then it is placed in nonlabelled medium of the same concentration. In order to estimate efflux uncomplicated by plant reabsorption of the isotope, the efflux medium is replaced with fresh, nonlabelled solution at regular intervals. From this time series of measurements, a pattern of efflux can be obtained (Fig. 3) in which apparent free space efflux can be separated from cytoplasmic and vacuolar effluxes by its higher rate (37,48, see chap. 6 of 29). Knowing the time required to saturate the apparent free space one can preincubate the experimental plants for this length of time (e.g. 5 min). Alternatively, one could disregard the uptake during the first 5 minutes in the nutrient disappearance studies. This choice is less desirable because if the apparent free space is relatively large, then the concentration of the limiting nutrient at the lowest concentrations used in the experiment may be lowered considerably before the uptake component of interest is observed.

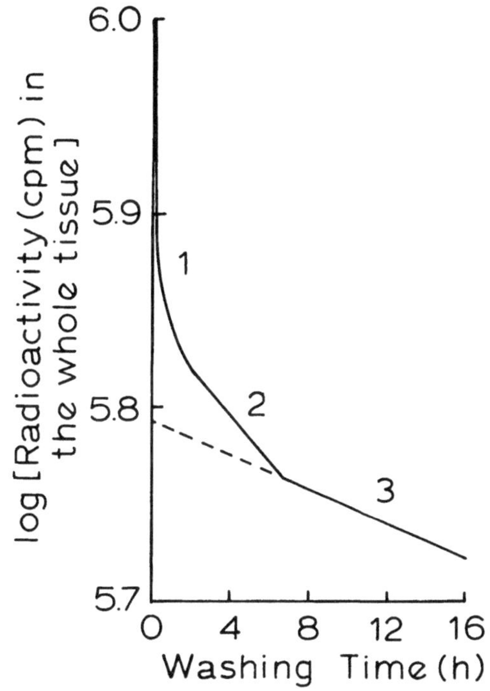

Fig. 3: Hypothetical time series showing efflux of isotope from plant tissue as a function of wash time. Region 1 represents the apparent free space, region 2 the cytoplasm and region 3 the vacuole. Radioactively labelled plants are placed in nonlabelled medium and the amount of label appearing in the wash medium is measured (see text for details).

Further tests to determine if uptake is active or passive, include the use of metabolic inhibitors, uptake determinations at different temperatures and different substrate concentrations. These tests have been used to characterize the active uptake of leucine by the brown alga, Giffordia mitchellae (37) and this research offers a good example to follow.

In both types of uptake experiments it is necessary to stir the medium to prevent a small zone of nutrient depletion from forming around the thallus. If this zone is allowed to develop, then the actual uptake rate is reduced because the nutrient ion must diffuse across this zone, giving rise to diffusion-limited uptake rates (33). In the kelp, Macrocystis pyrifera, Wheeler (52) found that water speeds of approximately 4 cm s^{-1} were necessary for maximum uptake rates. Mixing may be achieved by aerating the medium or by stirring with a magnetic bar. In the former case for uptake of ammonium at low concentrations, the air should be "scrubbed" free of ammonia by passing it through 1N $ZnCl_2$ and then through distilled water, to prevent ammonia in the air from being added during the aeration process. If a stirring bar is used, it should be placed inside some kind of mesh cage to prevent it from injuring the thallus, or the thallus may be suspended above the stirring bar.

Each of the two basic designs for nonisotopic nutrient uptake rate measurements has several advantages and disadvantages (46). The perturbation method allows one to observe how the pattern of uptake rate changes with time. This has been shown to be important in phytoplankton (6,10) and in higher plants (26,29). This experimental design could simulate some environmental situations that may be encountered by a macrophyte. For example, large amounts of nitrate may be brought into the euphotic zone by storm activity (49) or when an intrusion of high salinity bottom water replaces lower salinity interstitial water in the sediments, releasing large amounts of nutrients (40). The field response to such a large nutrient change can only be assessed with the perturbation experiment. The disadvantage of this method is that only one or a few plants are used and it is difficult to assess whether these plants are typical of the field population under study. Another disadvantage is that the nutrient past history is changing during the experiment (i.e. the plant continues to overcome its nutrient debt). This results in reduced uptake rate later in the experiment possibly due to a process such as feedback inhibition (6,26). This is not a serious disadvantage for this type of experiment, however, because such a process also occurs in the field.

The multiple container - constant incubation time experiment is recommended for determining the nutrient uptake parameters, V_{max} and K_s. Since a new or different plant is used for each concentration, and the incubation time is constant, the past history is the same for each uptake experiment, resulting in no particular bias for a certain set of measurements. An additional advantage of this type of experiment is that many different plants are used, which gives rise to more variability in the uptake rate data (44), but actually provides a better assessment of the population response since interplant variability is taken into account. To reduce the variability in the data, many more uptake measurements are required; each concentration should be run in triplicate at least.

In most studies it is desirable to obtain both the uptake pattern and the uptake kinetic parameters V_{max} and K_s for the population by incorporating interplant variability into the latter measurements. Therefore, it is recommended that both types of uptake experiments be carried out if possible.

II Biological considerations

1. Prehistory or preconditioning. In order to properly interpret uptake measurements, the preconditioning or prehistory of the plant in the field should be known. This can be accomplished by frequently and regularly monitoring the ambient limiting concentrations for at least several weeks in advance of the experiment, preferably throughout the depth range encountered by the plants. In the case of nitrogen, the limiting nutrient would include measuring at least nitrate and ammonium concentrations, but dissolved organic nitrogen may also be important. In an eleven year study of the English Channel, Butler et al. (1) showed that while nitrate and ammonium decreased to very low levels during the summer months, dissolved organic nitrogen rose during this period to significant levels (approximately 6-8 µM), indicating that it may be an important available source of nitrogen for some species.

The prehistory of the plant in the field can also be assessed by measuring the concentration of various internal components of the plant which change as the external nutrient concentration changes. For nitrogen limitation, the most sensitive parameter appears to be the internal nitrogen (nitrate, ammonium or amino acids) pools, which can be determined by hot ethanol extraction (4). In studies of *Laminaria saccharina* the internal nitrate pool increased at external nitrate concentrations of about 10 µM (5) and in *Macrocystis pyrifera* changed significantly in a few days in response to low external concentrations (53).

The C/N ratio also reflects the nutrient prehistory. In some continuous culture experiments with two red algae, DeBoer et al. (7) showed that C/N ratios >10 indicated nitrogen limitation and the ratio increased as the degree of limitation increased. However, some caution should be exercised in interpreting C/N ratios, because in estuarine areas nitrogen constituents have been reported to increase as salinity decreases (30) and the ratio may also vary with the type of alga (31); the Phaeophyta had higher ratios than either the Chlorophyta or Rhodophyta.

Another condition that must be considered in nutrient uptake experiments is the degree of desiccation of the alga. Recently, it has been shown that when some intertidal algae are desiccated upon emergence (as during the daytime in the summer at low tide) and then re-submerged, the subsequent uptake rate is enhanced due to a desiccation effect (43). In the high intertidal alga, *Fucus distichus*, the uptake rate was enhanced by a factor of approximately two when the alga was desiccated by 30%. For low intertidal and subtidal algae there was little or no enhancement of nutrient uptake upon desiccation. For a standard condition for interspecies comparisons, it is recommended that all species be completely hydrated before the beginning of the experiment.

2. Variation between plant parts. Uptake determinations on whole thalli are preferable for ecological measurements, but if the thallus is too large then a consideration of which portion of the thallus to use becomes important. Meristematic regions of the thallus are generally the most active metabolically. Therefore, if tissue from only this part is used for uptake measurements, extrapolation to whole plant uptake will result in an overestimation.

Cutting of the thallus to produce tissue segments may alter the uptake rate due to a wounding response resulting in increased respiration (23), or elimination of the transport system or active sink region (34). A comparison of excised sections of *Macrocystis* tissue with whole blades, showed a decrease in uptake rate as high as 80% by the cut sections (50). In addition, freshly cut discs of *Laminaria longicruris* have been reported to release brownish-colored substances and mucilage, and the water had to be changed before the experiment began (21).

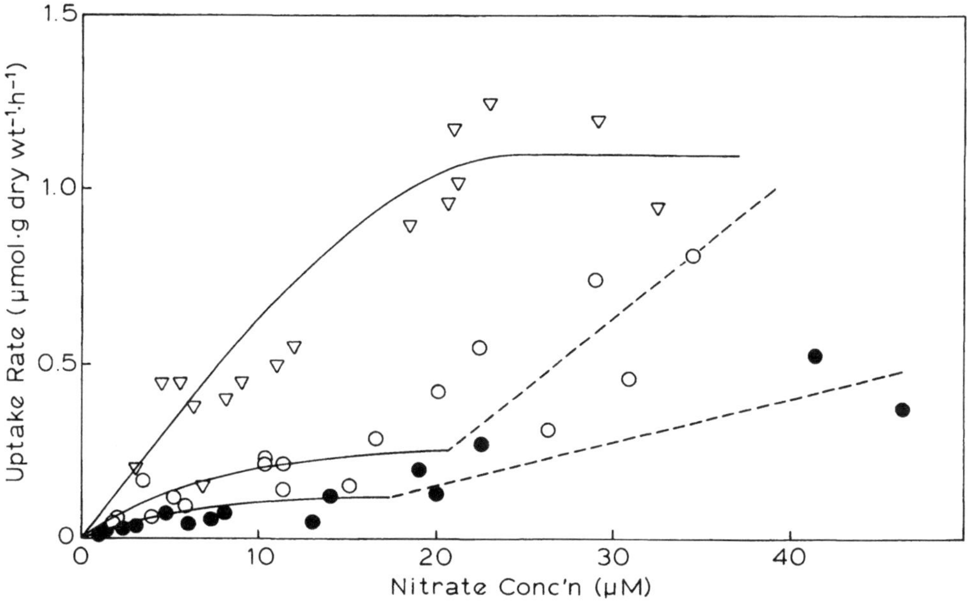

Fig. 4: Nitrate uptake rate for the first (Δ), second (o), and third (●) year whole plants of Laminaria groenlandica as a function of nitrate concentration. Rates were determined for whole plants except for third year plants where a longitudinal strip from the blade was used. The plants were incubated at an irradiance of 50 $\mu E\ m^{-2}\ s^{-1}$ for 1 h at 13 C in filtered and enriched sea water to which different concentrations of nitrate were added. About 5 g l^{-1} of plant tissue was placed in a large plexiglass tank and the sea water was frequently stirred during the incubation. Disappearance of nitrate from the medium (see reference 41 for method) was used to calculate the uptake rate. Data points represent a single rate measurement and curves were fitted visually.

3. Variation as a function of plant age. Many seaweeds are perennial, creating natural populations consisting of several age classes. Any extrapolation of nutrient studies to field conditions must recognize the possibility of different uptake kinetics for the resident age classes.

Our preliminary studies on the nitrate uptake kinetics of three year classes of Laminaria groenlandica have demonstrated a decreased uptake with increasing age (Fig. 4). Further, the shape of the uptake curve changed from the classical rectangular hyperbola for the first year plants, to a curve with enhanced uptake at higher nitrate concentrations for second and third year plants.

The opposite trend of uptake rate with age of tissue has been observed for Macrocystis by Schmitz and Srivastava (38). They found that punched discs from mature blades took up more phosphate than similar discs from young meristematic regions. In other experiments, they showed that the mature regions took up and transported the phosphorus to sink regions (meristems).

4. Interspecific variation. Comparative studies by Gagné and Mann (15) of Laminaria longicruris populations growing in areas of constant vs. seasonal nitrogen availability, indicated a possible different genetic response to these different nutrient regimes. This finding indicates the necessity of caution in extrapolating results obtained for a single population. To our knowledge there are no studies elucidating the degree of variability in nutrient uptake within populations.

Summary

Kelp biology has been expanding rapidly in the last few years. Physiological aspects such as nutrient uptake have been seen to be important in understanding their field biology. Frequently these physiological uptake experiments have been conducted by field ecologists who have insufficient exposure to various phytoplankton uptake methods to be able to decide on the appropriate method for their specific situation. We hope that this review will be useful for ecologists trying to choose the most appropriate procedures for their nutrient uptake experiments with macrophytes.

Acknowledgments

This research was supported by the Natural Sciences and Engineering Research Council of Canada. We thank Drs. W.N. Wheeler and D.H. Turpin, and T.E. Thomas for critically reviewing this manuscript. Technical assistance was provided by Ms. K. Lloyd.

References

1. Butler, E.I., Knox, S., Liddicoat, M.I.: J. mar. biol. Ass. UK 59, 239-250 (1979).

2. Caperon, J., Meyer, J.: Deep-Sea Res. 19, 619-632 (1972).

3. Chapman, A.R.O.: In Handbook of Phycological Methods: Culture Methods and Growth Measurements, (J.R. Stein, ed.), Cambridge University Press, London, pp. 88-104 1973

4. Chapman, A.R.O., Craigie, J.S.: Mar. Biol. 40, 197-205 (1977).

5. Chapman, A.R.O., Markham, J.W., Lüning, K.: J. Phycol. 14, 196-198 (1978).

6. Conway, H.L., Harrison, P.J., Davis, C.O.: Mar. Biol. 35, 187-199 (1976).

7. DeBoer, J.A., Guigli, H.J., Isreal, T.L., D'Elia, C.F.: J. Phycol. 14, 261-266 (1978).

8. DeBoer, J.A.: In The Biology of Seaweeds, (Lobban, C.S., Wynne, M.J., eds.), Blackwell, Oxford, in press

9. D'Elia, C.F., DeBoer, J.A.: J. Phycol. 14, 266-272 (1978).

10. DeManche, J.M., Curl Jr., H.C., Lundy, D.W., Donaghay, P.L.: Mar. Biol. 53, 323-333 (1979).

11. Dowd, J.E., Riggs, D.S.: J. Biol. Chem. 240, 863-869 (1965).

12. Eppley, R.W., Blinks, L.R.: Plant Physiol. 32, 63-64 (1957).

13. Epstein, E.: Mineral Nutrition of Plants: Principles and Perspectives, John Wiley and Sons, New York, 1972

14. Epstein, E.: In Encyclopedia of Plant Physiology, New Series, Vol. 2. Transport in Plants II., Part A, Cells, (U. Lüttge and M.G. Pitman, eds.), Springer Verlag, New York pp. 72-94 1976

15. Gagné, J.A., Mann, K.H.: Int. Seaweed Symp. 10, Abst. #B36 (1980).

16. Gessner, F., Hammer, L.: Mar. Biol. 1, 88-91 (1968).

17. Haines, K.C., Wheeler, P.A.: J. Phycol. 14, 319-324 (1978).

18. Hammer, L.: Mar. Biol. 4, 136-138 (1969).

19. Hanisak, M.D., Harlin, M.M.: J. Phycol. 14, 450-454 (1978).

20. Harlin, M.M.: Aquaculture 15, 373-376 (1978).

21. Harlin, M.M., Craigie, J.S.: J. Phycol. 14, 464-467 (1978).

22. Harrison, P.J., Conway, H.L., Holmes, R.W., Davis, C.O.: Mar. Biol. 43, 19-31 (1977).

23. Hatcher, B.G.: Mar. Biol. 43, 381-385 (1977).

24. Healey, F.P.: Microb. Ecol. 5, 281-286 (1980).

25. Johnston, C.S.: Int. Rev. ges. Hydrobiol. 54, 473-490 (1969).

26. Lefevre, D.D.: The Regulation of Phosphate Uptake by Intact Barley

Plants. M.Sc. Thesis, University of British Columbia, Vancouver, 1980

27. Lobban, C.S.: J. Phycol. 14, 178-182 (1978).

28. Lüning, K., Neushul, M.: Mar. Biol. 45, 297-309 (1978).

29. Lüttge, U. Higinbotham, N.: Transport in Plants. Springer Verlag, New York, 1979

30. Munda, I.M.: Aquatic Bot. 4, 347-351 (1978).

31. Niell, F.X.: Bot. Mar. 19, 347-350 (1976).

32. Nissen, P.: Physiol. Plant. 28, 304-316 (1973).

33. Pasciak, W.J., Gavis, J.: Limnol. Oceanogr. 19, 881-888 (1974).

34. Penot, M., Penot, M.: Z. Pflanzenphysiol. 95, 265-273 (1979).

35. Provasoli, L.: In Cultures and Collections of Algae, Proceedings U.S. - Japan Conference (A. Watanabe, A. Hattori, eds.), Hakone, Jap. Soc. Plant Physiol., Tokyo, pp. 63-75 1968

36. Ryther, J.H., Dunstan, W.M.: Science 171, 1008-1013 (1971).

37. Schmitz, K., Riffarth, W.: Z. Pflanzenphysiol. 96, 311-324 (1980).

38. Schmitz, K., Srivastava, L.M.: Plant Physiol. 63, 1003-1009 (1979).

39. Schoenbeck, M.W., Norton, T.A.: Bot. Mar. 22, 233-236 (1979).

40. Smetacek, V., von Bodungen, B., von Brodsel, K., Zeitschel, B.: Mar. Biol. 34, 373-378 (1976).

41. Strickland, J.D.H., Parsons, T.R.: A Practical Handbook of Seawater Analysis, Fish. Res. Board Canada, Bull. 167 (2nd Ed.), 310 pp. 1972

42. Takahashi, M., Seibert, D.L., Thomas, W.H.: Deep-Sea Res. 24, 775-780 (1977).

43. Thomas, T.E., Turpin, D.H.: Bot. Mar. 23, 479-481 (1980).

44. Topinka, J.A.: J. Phycol. 14, 241-247 (1978).

45. Towle, D.W., Pearse, J.S.: Limnol. Oceanogr. 18, 155-159 (1973).

46. Turpin, D.H.: Processes in Nutrient Based Phytoplankton Ecology, Ph.D. Thesis, University of British Columbia, Vancouver, 1980

47. Waite, T., Mitchell, R.: Bot. Mar. 15, 151-156 (1972).

48. Walker, N.A., Pitman, M.G.: In Encyclopedia of Plant Physiology, New Series, Vol. 2, Transport in Plants II, Part B, Tissues and Organs, (U. Lüttge and M.G. Pitman, eds.), Springer Verlag, New York, pp. 93-126 (1976).

49. Walsh, J.J., Whitledge, T.E., Kelley, J.C., Huntsman, S.A., Pillsbury, R.D.: Limnol. Oceanogr. 22, 264-280 (1977).

50. Wheeler, P.A.: J. Phycol. 15, 12-17 (1979).

51. Wheeler, P.A., North, B.B., Stephens, G.C.: Limnol. Oceanogr. 19, 249-259 (1974).

52. Wheeler, W.N.: Mar. Biol. 56, 103-110 (1980).

53. Wheeler, W.N.: In Synthetic and Degradative Processes in Marine Macrophytes. (L.M. Srivastava, ed.) Walter de Gruyter, Berlin, 1981

DISCUSSION

BIDWELL: Were the curves in your figure 4 done by mathematical analysis, or by a computer or did you draw them by eye?

HARRISON: They were drawn by eye. More work needs to be done, especially using higher concentrations, before these data become meaningful.

LÜNING: Could it be that this enormous difference in nitrate uptake by different age group plants is due to completely different growth activity; that is, that the first year plants are much more fast growing than the second and third year plants?

DRUEHL: The first two years' plants appeared to have the same growth rate on both surface area and fresh weight basis, in the third year the growth rate was greatly increased.

LÜNING: Increased!

DRUEHL: That is correct.

McLACHLAN: I am still curious about this difference between the first, second and third year. Could it be that the second and third year plants have more inactive (mature, Ed.) tissue and supply sufficient amounts of nitrogen to the growing portions of the plants and that perhaps uptake curves are rather similar?

HARRISON: That is a good idea and it should be done -- that is, to take active tissues from each one of these age classes and compare uptake in active tissues.

SRIVASTAVA: Might not the free space component itself be different in one year *vs.* two year material?

HARRISON: It probably is, but we could not detect any change in the uptake rate over a 2 h period, sampling every 15 min.

NITROGEN NUTRITION OF <u>MACROCYSTIS</u>

W. N. Wheeler[1]
Marine Science Institute, University of California, Santa Barbara, USA

Introduction

Early in this century, many of the larger kelps were harvested for potash. Later, other chemicals, iodine and alginic acid, were found to be present in the brown algae in commercially significant quantities. The harvesting and subsequent processing of kelps thus became the basis for an industry, which is now world-wide (5). Recently, interest has been diverted toward the utilization of brown algae, especially Macrocystis, as a renewable source of biomass for energy production. Economically viable quantities of the plant, however, can only be achieved under intensive cultivation, and nutrient fertilization of large kelp beds is likely to be an expensive necessity. Answers to questions of nitrogen nutrition of the large kelps, therefore, are paramount to a successful cultivation effort.

Heretofore, little work has been done outside of growing laboratory-reared sporophytes in defined media and there is hardly any information on the nitrogen nutrition of adult sporophytes in the field. Since any cultivation effort must take place in the sea, with adult plants, a knowledge of the nutrient ecology of the adult sporophyte is fundamental to any fertilization strategy. The emphasis of this paper, therefore, is on the adult sporophyte and its response to its nitrogen environment.

Although the data presented are predominently my own work, I have tried to set this paper within the context of the available literature on the nitrogen physiology of Macrocystis and other marine macroalgae.

[1] Present Address: Department of Biological Sciences, Simon Fraser University, Burnaby, B.C. V5A 1S6, Canada

Synthetic and Degradative Processes in Marine Macrophytes

Materials and Methods

Whole surface and subsurface fronds of Macrocystis pyrifera were collected from kelp forests near the University of California, Santa Barbara, USA. They were transported to the laboratory within 1 h, and held in outdoor tanks with flowing seawater for a maximum of 36 h before analysis. Disks (8 cm or 1.3 cm in diameter) were cut from blades and cleaned of visible epiphytes. The cut disks were then preincubated in experimental media for 1 h. Preincubation allowed the removal of unwanted wound exudates, gave time for the filling of free space within the tissue with the experimental medium in use, and may have lessened the effect of any wound reaction on the uptake process.

Incubation media consisted of nitrogen-depleted seawater with added, known concentrations of KNO_3 and/or NH_4Cl. Tissue disks were added to jars (1 liter or 20 ml capacity) with tissue to medium volume ratio of 3g liter^{-1} or less. Most experiments were run for 1 h, and the concentrations of nitrogen in the water before and after the incubation were measured. Earlier experiments involving brief treatments of disks with a variety of disinfectants, including formaldehyde, bleach, and various antibiotics, showed that uptake rates were unaffected by bacterial contaminants over the experimental period of 1 h.

Irradiance was provided by a tungsten-iodide lamp (GE Quartzline Q150013/C4) which gave a maximum irradiance of 300 μE m^{-2} s^{-1}. Temperature was held constant at 15 C ($\pm$ 0.5 C) using either a water bath or cold room. Water motion in the incubation jar was provided with synchronized magnetic stirrers (see 29,30) or with a shaker. Nitrate and ammonium determinations were made according to Strickland and Parsons (25). Tissue nitrate was determined according to Chapman and Craigie (3).

Results and Discussion

Adult plants of Macrocystis pyrifera near Santa Barbara can have as many as 100 fronds. Each frond may grow to a length of 20 m or more and produce hundreds of blades. Wheeler (29) has shown that selective sampling of blade tissue along a number of fronds of different ages gives a general picture of frond and plant development. There is an age gradient among the blades of a frond from the apical meristem to the base (holdfast). As in some terrestial grasses (21), these ontogenetic differences are pronounced and appear to overshadow irradiance effects on pigments and photosynthesis. It is, thus, imperative to first focus on the ontogenetic variability among blades.

Nitrate uptake

Ontogenetic effects. The blades of *M. pyrifera* were shown to have only a slight photosynthetic gradient on a surface area basis from the proximal to the distal end by Clendenning (6). A similar shallow gradient seems to prevail for nitrate uptake along the length of a blade (Fig. 1).

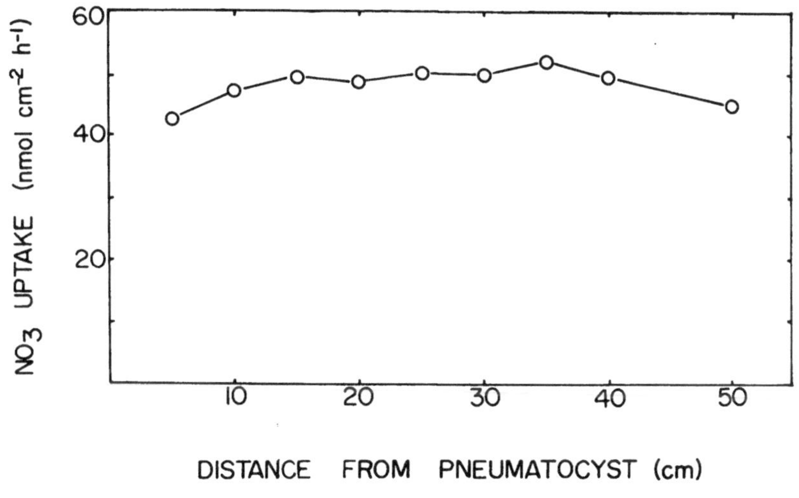

Fig. 1: Nitrate uptake of blade disks along a mature blade of *Macrocystis pyrifera*.

Blade disks sampled from each meter segment of subsurface fronds yield uptake rates for NO_3 (Fig. 2), which seem to follow the photosynthetic pattern described for blades along the length of a frond (29). In general, the disks from the embryonic blades near the meristem showed low uptake rates when measured under optimal uptake conditions (300 $\mu E\ m^{-2}\ s^{-1}$; 25 $\mu M\ NO_3$; $\geq$ 6 cm s^{-1} water speeds; 15 C). The disks from the

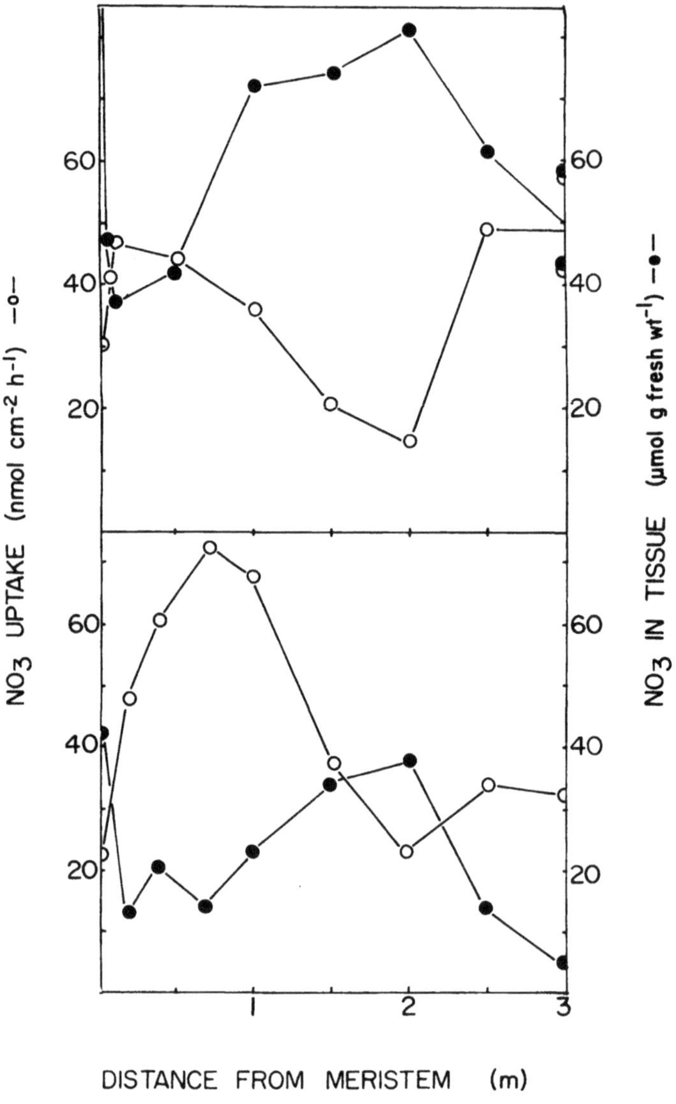

Fig. 2: Distribution of nitrate uptake and concentration of nitrate in the tissue along two juvenile fronds of Macrocystis pyrifera. Two small disks were cut from blades about 30 cm from the pneumatocyst at increasing distances from the apical meristem. Open circles represent disks used for uptake determinations; closed circles represent the nitrate concentrations of adjacent disks. The top figure shows a frond with high tissue nitrate, the bottom figure a frond where nitrate is low. The points at 3 m represent data for sporophylls.

mature blades showed the highest uptake rates. Senescing blades also showed relatively low uptake rates.

It has been hypothesized (16) that sporophylls, because of their position below the thermocline, may be important centers of nitrate uptake. These blades do not show signs of senescence so noticeable in other blades, and therefore persist longer. There is, however, no evidence from my data that their uptake rates are any higher than those of vegetative blades at corresponding depths (Fig. 2).

Concentration of nitrate in the tissue may also play a role in the nitrate uptake capacity of that tissue. Figure 2 shows that if the nitrate concentration in the tissue is high the uptake rates may be correspondingly low (upper graph) and _vice versa_ (lower graph; see also section on storage). The ability of a particular tissue to take up nitrate, thus, might be the result of a combination of the past growth history and feedback inhibition. Experiments to separate the two effects are now under way.

Light. A high photon flux density may enhance the uptake of nitrate by mature blades in _M. pyrifera_, although the data presented are not statistically significant (Fig. 3). These data, however, do indicate a substantial uptake of nitrate at 0 $\mu E\ m^{-2}\ s^{-1}$ (dark).

Water motion. Water motion, delivered either as a defined uni-directional current in a water tunnel, or as a general turbulence in incubation chambers, produces a significant effect on the uptake of nitrate by mature blade disks of _M. pyrifera_ (Fig. 4). Nitrate uptake rates in one set of experiments rose from 9 to 44 nmol $cm^{-2}\ h^{-1}$, an enhancement of 488%, as water motion increased from 0 to 3 $cm\ s^{-1}$.

Concentration of nitrate. Under saturating water motion (6 $cm\ s^{-1}$) and photon flux densities (300 $\mu E\ m^{-2}\ s^{-1}$), disks from mature blades of _M. pyrifera_ respond to changes in ambient nitrate concentration in a hyperbolic manner (Fig. 5). Although the form is Michaelis-Menten, this in no way implies a knowledge of the mechanisms involved. Using the terminology of Michaelis-Menten kinetics allows comparison of uptake curves between different species. The V_{max} is the maximum uptake

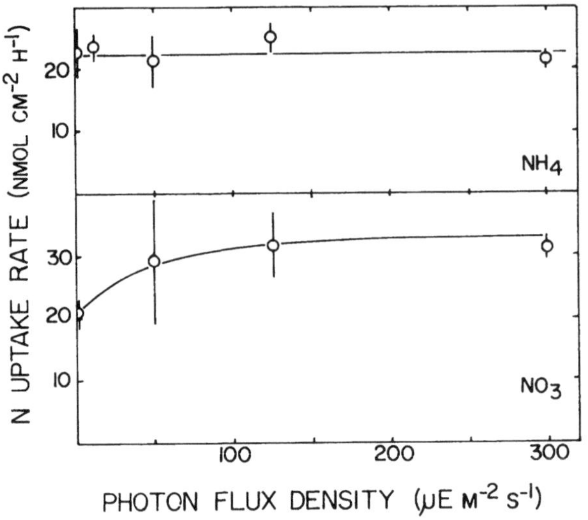

Fig. 3: Nitrate and ammonium uptake rates vs. irradiance for Macrocystis pyrifera mature blade disks. Top curve shows uptake of ammonium; NH_4 concentration in the incubation medium 9 μM. Bottom curve shows uptake of nitrate; NO_3 concentration in the medium was 9 μM. Sample size, 10 blade disks; vertical bars represent S.E..

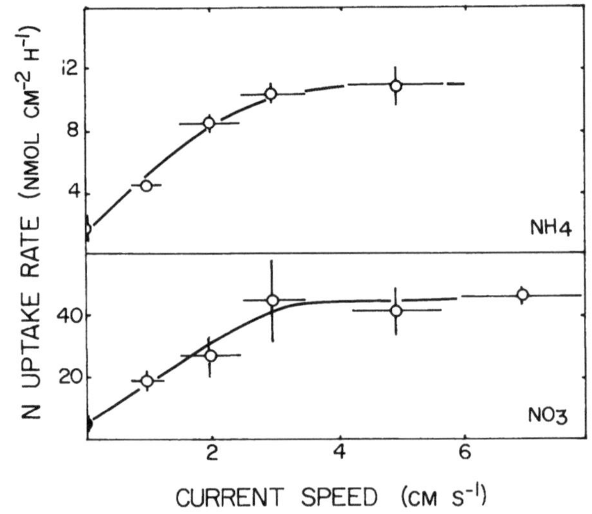

Fig. 4: Nitrate and ammonium uptake rates vs. water motion for Macrocystis pyrifera mature blade disks. Top curve shows uptake of ammonium; NH_4 concentration in the incubation medium was 3.5 μM. Bottom curve shows uptake of nitrate; NO_3 concentration in the medium was 15 μM. Sample size, 10 blade disks; vertical bars represent S.E.. Horizontal bars represent S.E. of 50 water speed measurements (turbulent fluctuations).

rate, or the uptake rate at infinite ambient concentration of NO_3; the K_s is the ambient concentration at 1/2 the maximum uptake rate. For the mature blade of M. pyrifera, the V_{max} is approximately 75 nmol cm^{-2} h^{-1} the K_s is approximately 13 μM. For young sporophytes of M. pyrifera, Haines and Wheeler (11) calculated a V_{max} of 95 nmol cm^{-2} h^{-1} for nutrient-rich, deep-water-grown plants and 69 nmol cm^{-2} h^{-1} for nutrient-poor surface-water-grown plants. The K_s values varied accordingly, being 13.1 μM for deep water plants and 8.7 μM for the surface water plants. The above numbers and others have been converted to a fresh weight basis for comparison (Table I).

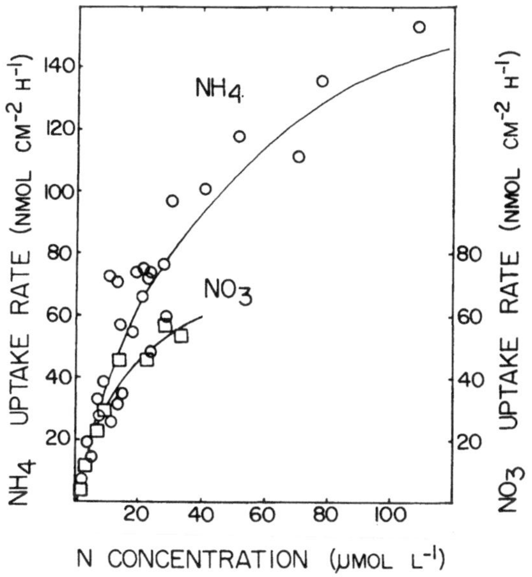

Fig. 5: Nitrate and ammonium uptake rates vs. concentration for Macrocystis pyrifera mature blade disks. Disks were incubated in varying ambient concentrations of NH_4 (circles) and NO_3 (squares). During incubation saturating conditions of irradiance (300 μE/m^2/sec) and water motion (6 cm/sec) were maintained.

Ammonium uptake

Ontogenetic effects. Wheeler (28) using methylamine as an analogue of the ammonium ion demonstrated that apical sections and juvenile sporophytes of M. pyrifera had much lower uptake rates than mature tissues. But, using

Table I: Nitrate uptake constants.

Alga	Temp (C)	Ks (µM)	Vmax (µmol g fr wt^{-1} h^{-1})	Ref
Codium fragile	6	1.9	0.28[a]	(12)
	12	3.8	0.60[a]	"
	18	6.9	1.09[a]	"
	24	7.7	0.96[a]	"
Fucus spiralis	5	5.6	3.48[b]	(27)
	10	6.7	4.39[b]	"
	15	7.8	4.96[b]	"
Laminaria longicruris	0	nd	1.19[c]	(13)
	12	nd	1.63[c]	"
Macrocystis pyrifera	16	13.1	3.05[d]	(11)
	16	8.7	2.25[e]	"
	15	13.0	3.43[f]	(here)
Hypnea musciformis	26	4.9	2.85	(11)
Gracilaria foliifera	20	2.5	0.97[a]	(8)
Neoagardhiella baileyi	20	2.4	1.16[a]	(8)

[a] Wet wt/dry wt = 10.
[b] Wet wt/area = 0.046.
[c] Wet wt/dry wt = 5.9.
[d] Young, deep water sporophytes.
[e] Young, surface water sporophytes.
[f] Mature blades.
nd Not determined.

cut disks from blades, apical sections were found to have higher uptake rates than whole blades.

Light. Variation in photon flux density appears to have no effect on the rate of ammonium uptake in mature blades of M. pyrifera (Fig. 3).

Water motion. Water motion, either as a uni-directional current in a water tunnel or as turbulent flow in an incubation chamber increases uptake of ammonium by M. pyrifera blade disks (Fig. 4). For example, under an ambient concentration of 3.5 µM ammonium, the uptake rates increased from 2 to 11 nmol cm^{-2} h^{-1} when water speeds increased from 0 to 5 cm s^{-1}.

Concentration of ammonium. Ammonium uptake kinetics appear complex. For

Table II: Ammonium uptake constants.

Alga	Temp (C)	Ks (µM)	Vmax (μmol g fr wt^{-1} h^{-1})	Ref
Codium fragile	6	1.47	1.30[a]	(12)
	18	1.56	2.78[a]	"
	30	2.1	2.15[a]	"
Fucus spiralis	5	6.4	3.9[b]	(27)
	10	6.4	5.6[b]	"
	15	9.6	7.7[b]	"
Macrocystis pyrifera	16	5.3	2.38[c]	(11)
Macrocystis pyrifera	8	50	2.36[d]	(28)
Macrocystis pyrifera	15	33	4.17[e]	(here)
Gracilaria foliifera	20	1.6	2.38[a]	(8)
Hypnea musciformis	26	16.6	not saturatable	(11)
Neoagardhiella baileyi	20	4.9	1.19[a]	(8)

[a] Wet wt/dry wt = 10.
[b] Wet wt/area = 0.046.
[c] Embryonic sporophytes, below 22 µM.
[d] Methylamine uptake in mature blades.
[e] Mature sporophyte blades; area/wtwt = 45.8.

juvenile M. pyrifera Haines and Wheeler (11) reported a sharp discontinuity in the uptake curve above 20 µM ambient NH_4 and interpreted that as being indicative of a 2 component uptake system. One system was believed to be composed of a high affinity (or low concentration) uptake mechanism, the other of a low affinity (or high concentration) mechanism. By contrast, Wheeler (28), using methylamine on mature blade disks showed saturation slightly above 100 µM with a K_s of about 50 µM. She claimed that because these numbers were 10 times higher than corresponding values for ammonium using juvenile sporophytes, the affinity of the carrier system for methylamine was 1/10 that of ammonium. However, the data presented here for ammonium using mature blade disks agree well with the methylamine data of Wheeler (28). The difference in uptake by young sporophylls and mature tissues noted by Wheeler (28) may, therefore, be age related and not be due to a lesser affinity for methylamine by the carrier system. One can only wonder why the saturation levels are so high when measured concentrations of ammonium in coastal waters rarely exceed

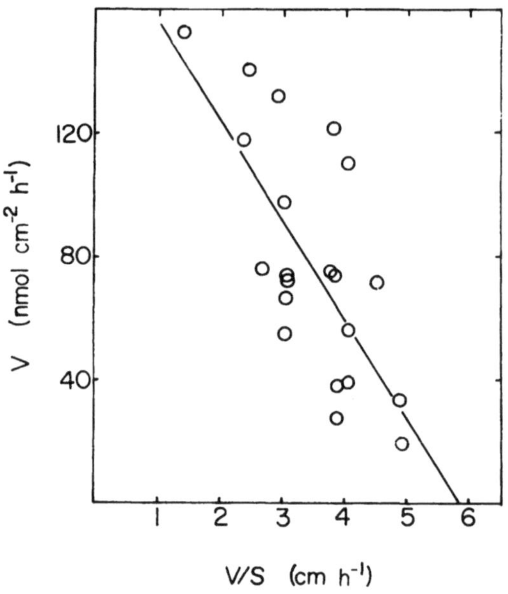

Fig. 6: Woolf-Hofstee plot of ammonium uptake between 10^{-6} and 10^{-4} M. Data from figure 5 of mature blades of Macrocystis pyrifera are presented here. r= 0.727, n=21, y= -32.5x + 191.

5 µM. Table II gives V_{max} and K_s data for ammonium uptake for a number of different marine algae. Only Macrocystis and perhaps Hypnea appear to have such low affinity uptake mechanisms.

Because the uptake of methylamine or ammonium has a pH optimum (pH 8.5), is temperature dependent, and is competitively inhibited by analogues, Wheeler (28) has concluded that the uptake of ammonium (or methylamine) is active and brought about by an "ammonium permease." The Woolf-Hofstee plot of ammonium uptake in mature blade disks shown in figure 6 also indicates an active uptake of ammonium.

In a number of other plant systems, including phytoplankton and higher plants, ammonium uptake is known to block the simultaneous uptake of nitrate, presumably by a feed-back inhibition. In many macroalgal systems, however, both nitrate and ammonium appear to be taken up simultaneously. Ammonium may be taken up preferentially as in Gracilaria (8) or it may have no effect whatsoever on nitrate uptake, as is the case with M. pyrifera (11).

Organic sources of nitrogen

Urea is a common source of organic N used as fertilizer for higher plants. North (22) has shown that gametophytes of M. pyrifera show positive growth in a medium containing less than 0.01 M urea as the sole nitrogen source. Direct uptake studies have not been done. Urea, supplied at concentrations of less than 10 ppm, has been shown to enhance the growth rate of Laminaria angustata and L. religiosa (14). Recently, Kirk and Kirk (18) have shown that several members of the Chlorophyta are able to take up and utilize amino acids. Schmitz and Riffarth (24) have shown active uptake and assimilation of a number of amino acids by Giffordia. They also showed that leucine had the highest uptake rate with a K_s of 30 to 120 mM and a V_{max} of 3 to 3.8 µmol 100 g dry wt^{-1} h^{-1}.

Storage

Chapman and Craigie (3) demonstrated that Laminaria longicruris could store nitrate in concentrations as high as 28,000 times over ambient. This nitrate would remain in the tissues for as long as 2 months after nitrate had dissappeared from the surrounding water. Buggeln (2) also found that Alaria could take up and store nitrate as much as 3000 times over ambient, but, contrary to Laminaria, Alaria stored nitrate only for periods of a few days.

Because a surplus of nitrate appears in the tissues of Macrocystis pyrifera, it appears that the uptake of nitrate is faster than its reduction (Fig. 2). Concentration of nitrate in M. pyrifera tissues can be as high as 120 µmol g fresh wt^{-1} although concentrations from 30-60 µmol g fresh wt^{-1} are more common. These values are very similar to those recorded for Laminaria saccharina in culture (4), and for natural populations of Alaria (2) and Laminaria longicruris (3).

Assimilation

My unpublished observations indicate that Macrocystis pyrifera, like Alaria, can reduce up to 50 µmol g fresh wt^{-1} of stored nitrate in 7 days, an assimilation rate of approximately 0.3 µmol g fresh wt^{-1} h^{-1}. The sporophytes of Laminaria assimilated 50 µmol of stored nitrate per gram of fresh tissue over 2 months, a reduction rate of 0.034 µmol g fresh wt^{-1} h^{-1}, which is almost ten times slower than the rate estimated for M. pyrifera. These values for the estimated loss rate of nitrate from the tissue and by implication nitrate reductase activity are not as high as those recorded for higher plants (see 20). Nitrate reductase activity in terrestrial plants grown with and without nitrate fertilizer range from 0.1 to 14.6 µmol nitrite g fresh wt^{-1} h^{-1} formed; the computed values of 0.034 to 0.3 µmol g fresh wt^{-1} h^{-1} for kelps thus are on the low end.

Nitrate reductase has been demonstrated in Undaria and Alaria (26). Further, the data of Takagi and Murata (26) suggest that the enzyme has its highest activity in the stipes and holdfast regions, followed by the transition region, the sporophyll and then the blade. Some work with young, cultivated sporophytes of Laminaria saccharina indicates that in vitro nitrate reductase activity, although present, is weak (unpublished data). Nitrate reductase has been characterized from the red alga, Porphyra (1). In Porphyra, the NADH-dependent reduction has an apparent K_s of 50 µM nitrate.

The assimilation of nitrate through nitrate reductase produces nitrite and finally ammonium. From ammonium, the assimilation of nitrogen appears to go predominantly through the glutamine synthetase pathway (15). High activities of glutamine synthetase have been measured in Laminaria saccharina (unpublished data) and Giffordia (17). Using $^{15}NO_3$, Haxen (15) demonstrated that in Macrocystis angustifolia the major portion of reduced N was incorporated into glutamate via the glutamine synthetase pathway. Assimilation of nitrogen through aspartate and alanine was very small. Feeding with ^{15}N-glutamine indicated that glutamate was the major acceptor. Inhibition of glutamine synthetase via methionine sulphoxamine stopped assimilation of nitrogen and produced increased levels of ammonium. Common amino acids found in the free state in the fronds and stipes of M. angustifolia from S. Africa were alanine (greatest in stipe), glutamine and glutamate (15).

Seasonal changes in nitrate concentration in seawater

In order to understand the nitrogen ecology of Macrocystis, it is important to understand the seasonal variability of nitrate in coastal waters. Typically in southern California, upwelling brings nitrate into the surface waters during the months of March, April and May. The nitrate dissappears from the surface waters, except in trace amounts, after June. In many cases, this amount does not climb again until the following March. For almost 8 months, then, Macrocystis sp. must get its nutrients from transient local mixing, land runoff, or utilize the slightly higher concentrations available below the thermocline, a region usually no greater than 1 m above the bottom.

Macrocystis pyrifera standing crops, as deduced from harvest records, have been shown to vary seasonally. Harger (10) calculated that half of the seasonal growth variability was due to seasonally variable upwelling and surface irradiance. Upwelling and water temperature were found to play the major roles in determining M. pyrifera growth in another study (19). Individual frond and plant growth rates of M. pyrifera were higher during the spring than in late fall which corresponds well with high and low nitrate concentrations in surface waters (7). Fertilizing M. pyrifera in situ with chemical inorganic nitrogen fertilizers or artifically upwelled "deep" water has resulted in increased growth rates (23). Inorganic nitrogen is probably the most limiting nutrient for growth of M. pyrifera in southern California. This is not true of M. pyrifera in central California (9), where substrate type, irradiance, and excess water motion (storms) seem to determine kelp growth and population dynamics.

Acknoledgements

I would like to thank D. Coon, P.J. Harrison, and J. Woessner for reading the manuscript. Parts of this work were completed in partial fulfillment of the degree of Doctor of Philosophy at University of California, Santa Barbara. Parts of the work were supported by the Department of Commerce, N.O.A.A. Office of Sea Grants - Grant No. 04-7-158-44121 and the State Resources Agency projects R/FA 10 and R/A 16A and National Science

Foundation Grants GA-27484 and ENG76-22720 to M. Neushul and A.C. Charters. Equipment and supplies were supplied by Neushul Mariculture Inc.

References

1. Araki,S., Ikawa T., Oohusa, T., Nisizawa, K.: Bull. Jap. Soc. Sci. Fish. 45, 919-924 (1979).

2. Buggeln, R.G.: J. Phycol. 14, 156-160 (1978).

3. Chapman, A.R.O., Craigie, J.S.: Mar. Biol. 40, 197-205 (1977).

4. Chapman, A.R.O., Markham, J.W., Lüning, K.: J. Phycol. 14, 195-198 (1978).

5. Chapman, V.G.: Seaweeds and their uses, Methuen, London 1950

6. Clendenning K.A.: Proc. Int. Seaweed Symp. 4, 55-65 (1963).

7. Coon, D.A.: Bot. Mar. 24, 19-27 (1981).

8. D'Elia, C.F., DeBoer, J.A.: J. Phycol. 14, 266-272 (1978).

9. Foster, M.S.: In Synthetic and degradative processes in marine macrophytes (L.M. Srivastava, ed.), Walter de Gruyter, Berlin - New York 1981

10. Harger, B.W.W.: Coastal Oceanography and hard substrate ecology in a Californian kelp forest. Ph.D. Thesis. University of California, Santa Barbara 1979

11. Haines, K.C., Wheeler, P.A.: J. Phycol. 14, 319-324 (1978).

12. Hanisack, M.D., Harlin, M.M.: J. Phycol. 14, 450-454 (1978).

13. Harlin, M.M., Craigie, J.S.: J. Phycol. 14, 464-467 (1978).

14. Hasagawa, Y., Komaki, S., Honma, K.: Hokkaido Fish. Bull. 19, 165-187 (1962).

15. Haxen, P.C.: The major pathway of nitrogen assimilation in the marine kelp, *Macrocystis angustifolia* Bory. B.S. Honors Thesis, University of Cape Town, Cape Town, South Africa 1978

16. Jackson, G.A.: Nutrients and productivity of the giant kelp, *Macrocystis pyrifera*, in the near-shore. Ph.D. Thesis, California Institute of Technology, Pasadena, California 1975

17. Kiefer, H.: Untersuchungen zur Characterizierung und physiologischen Regulation der Nitratreductase und des Glutaminsynthetase bei der marinen Braunalge *Giffordia mitchellae*. Diplomarbeit, Universität Köln, FRG 1980.

18. Kirk, D.L., Kirk, M.M.: J. Phycol. 14, 198-203 (1978).

19. Kirkwood, P.D.: Seasonal patterns in the growth of the giant kelp,

Macrocystis pyrifera. Ph.D. Thesis, California Institute of Technology, Pasadena, California 1977

20. Lee, J.A., Stewart, G.R.: Adv. Bot. Res. _6_, 1-43 (1978).
21. Ludlow, M.M., Wilson, G.L.: Aust. J. biol. Sci. _24_, 1077-1087 (1971).
22. North, W.J.: Kelp habitat improvement project. Annual report, July 1974 - June 1975
23. North, W.J., Gerard, V.A.: _In_ Synthetic and degradative processes in marine macrophytes (L.M. Srivastava, ed.), Walter de Gruyter, Berlin - New York 1981
24. Schmitz, K., Riffarth, W.: Z. Pflanzenphysiol. _96_, 311-324 (1980).
25. Strickland, J.D.H., Parsons, T.R.: Bull. Fish. Res. Board Canada _167_, 1-310 (1972).
26. Tagaki, M., Murata, K.: Bull. Fac. Fish. Hokkaido Univ. _6_, 25-28 (1955).
27. Topinka, J.A.: J. Phycol. _14_, 241-247 (1978).
28. Wheeler, P.A.: J. Phycol. _15_, 12-17 (1979).
29. Wheeler, W.N.: Mar. Biol. _56_, 97-102 (1980).
30. Wheeler, W.N.: Mar. Biol. _56_, 103-110 (1980).

DISCUSSION

WILLENBRINK: Did you do a time curve or the kinetics for the nitrate uptake?

WHEELER: Yes, although not shown, uptake of nitrate appears to be unsaturable for at least 12 h. The curves presented here were measured while the uptake was linear and nonsaturated.

BIDWELL: K_m properly should be applied to single stage enzymatic systems and I do not think you could honestly apply it here. For if you do, you are in great danger of drawing erroneous conclusions for a system such as uptake of an ion by an entire plant. For as soon as a little bit has been taken up that would affect the uptake rate. The diminution of rate which you thought was the attainment of the maximum velocity might in fact be due to removal or metabolism of the ion at the other side. The other thing I was concerned about was that your data for nitrate uptake (Fig. 5) do not show saturation, yet you calculate saturation values. I think that in both cases, nitrate and ammonium, you were pushing the data a bit far.

WHEELER: You are quite right about the definition of K_m. The same problem arises here as in the previous question. I have previously shown that the uptake in these kelp discs is linear over at least a 12 h period. Under these conditions, saturation of the internal pools is not found. Although in its strictest sense, the K_m can only be applied to enzyme systems, it gives us here a framework within which we can effectively compare interspecies differences in uptake behavior. As to your last comment, I would also agree perhaps there is not enough data to strictly calculate a saturation point. The main purpose of the study was, however, not to totally define the characteristics of the curve, but to look at uptake behavior over a natural range of concentrations.

ALBRIGHT: Do you think that some of the problems with uptake of materials may be caused by epiphytes, particularly bacteria and perhaps microalgae, because they can operate at extremely low concentrations with very high uptake rates.

WHEELER: I have done some work with antibiotics and so have Pat Wheeler and Marilyn Harlin. What we find is that over short periods, one or two h of incubation, the bacterial and microbial portion of the uptake is very small and can usually be neglected.

ALBRIGHT: In what range is it?

WHEELER: The highest rate found for microbial or other uptake is about 20% of the total after an 8 h incubation. For short incubation periods of one h or less, it is usually less than 10%.

SCHMITZ: Your measurements were made by noting the decrease of the ion in the medium. How much is the free space in the discs? Are all these data corrected for uptake in the free space?

WHEELER: All experiments were started after a one h preincubation period during which the plant discs recovered from wounding, filled free space and adapted to experimental conditions.

McLACHLAN: I am bothered by the uptake mechanism being proposed here, which seems to disregard the dissociation of ammonium in sea water, especially when concentrations of 100 to 200 μM are involved.

WHEELER: The ammonium uptake mechanism in Macrocystis is still under debate. There are two lines of evidence that suggest that the uptake may be active, but whether there are one or two separate systems is still under debate. The data above 100 μM simply extend measurements taken at lower concentrations, and add little to the explanation of the mechanism involved.

McLACHLAN: That still does not take into account the dissociation of ammonium in sea water.

ALBRIGHT: The use of antibiotics for inhibition of marine or other bacteria has come under extreme criticism. In many cases it just does not work.

McLACHLAN: Both the antibiotics are not effective against gram negative bacteria.

ALBRIGHT: Absolutely, and 75% of the bacteria isolated in sea water or almost any water are gram negative.

UPTAKE OF INORGANIC IONS AND THEIR LONG DISTANCE TRANSPORT IN FUCALES AND LAMINARIALES

J.Y. Floc'h

Laboratoire de Physiologie Végétale, Faculté des Sciences, Université de Bretagne Occidentale, F-29283-BREST. Cedex, France.

Introduction

As is true for land plants, marine macrophytes also require certain essential elements for their growth, differentiation and reproduction. However, in contrast to the land plants, which absorb inorganic ions from soil by their roots, the marine algae are able to take up their nutrients directly from the ambient sea water through all parts of their thalli. Most of the mineral elements which are known to occur in sea water have also been found in seaweeds. Moreover, it has been known from the earliest investigations that these plants have the ability to accumulate in their tissues some of the elements which are present in sea water only in trace amounts. As such, these plants have served as important commercial sources for I, K, and Na (for review see 1), and the mineral content of many of these species has been carefully determined (2,3). More recent studies have given comparative values on the distribution of the various mineral elements in different species (4-8), and in different morphological regions of the thallus, i.e. stipe and lamina (9-12). However, until recently, with a few exceptions (13,14), no detailed information was available on the variability in the mineral content of different regions and tissues of the larger algae. In the last ten years, various analyses of mineral elements have been performed in the Laminariales and Fucales, with regard to the age and function of different tissues (15-24), and from these studies it has become clear that the mineral composition of the larger algae is a function not only of direct uptake of elements from the sea water, but also of their long distance transport through the thallus.

Synthetic and Degradative Processes in Marine Macrophytes

If we exclude from consideration the vast literature dealing with permeability of giant algal cells, such as those of Nitella and Chara, to various ions it appears that much of the recent research on ion uptake by larger algae has been stimulated by one or more of the following considerations:

- mass cultivation of algae including use of fertilizers (25-28),
- growth regulation (29),
- effect of pollution on algal growth (30,31),
- monitoring changes in sea water (32-34),
- sewage cleaning (26),
- detoxification of animal feed (35),
- food chain (36), and
- human food(37).

In the present paper, I will deal with uptake and long distance transport of inorganic ions in the large brown algae in as far as these processes result in the accumulation of these ions in the thallus.

Uptake and Accumulation of Ions

Concentration factors

"Concentration factor" is defined as the ratio of the concentration of a mineral element in algal tissue to its concentration in sea water. Table I shows that concentration factors differ widely for different mineral elements, and for different species taken from the same habitat. They differ also for different habitats, and for different seasons (41,42).

Some of the highest concentration factors are known for Mn, PO_4, I ar NO_3. Other high concentration factors are known for Ti, Fe, Cr, Cu and Pb (10^4 - 10^6) (4,8,43).

Some authors have described an ascending order of concentration factors for Na < F < Ca < Mo < K < Co < Cu < As < Ni < Mn < Zn < I in various algae, including the Fucales and Laminariales; the concentration factors ranging from 1 or less for Na to 6 x 10^4 for I (9). Others have

Table I: Concentration factors (amount per unit fresh (f) or dry (d) weight of algae/ amount per unit sea water) for some ions whose uptake or translocation is reported in large algae.

Ions	Algae	Conc. Factor	Standard	Reference
Mn^{2+}	Fucus	$0.2 - 2.2 \times 10^5$	d	(38)
"	Sargassum	0.7×10^4	d	(8)
"	Laminaria	0.8×10^4	d	(8)
"	Laminaria	236	f	(39)
Zn^{2+}	Fucus	$1.7 - 2.5 \times 10^4$	d	(38)
"	Fucus	$0.6 - 1.1 \times 10^3$	f	(4)
"	Ascophyllum	1.4×10^3	f	(4)
"	Laminaria	$0.4 - 1 \times 10^3$	f	(4)
"	Laminaria	2.4×10^3	f	(39)
"	Laminaria	70 - 150	d	(8)
Ca^{2+}	Fucus	3 - 4.5	f	(40)
"	Laminaria	4.5 - 7	f	(40)
"	Laminaria	5 or less	f	(16)
"	various algae	1 - 12	d	(9)
Mg^{2+}	Laminaria	1.2 or less	f	(16)
K^+	various algae	10 - 30	d	(9)
"	Laminaria	48 or less	f	(16)
Na^+	various algae	1 or less	d	(9)
"	Laminaria	0.7 or less	f	(16)
Cl^-	Laminaria	less than 1	f	(16)
SO_4^{2-}	Laminaria	3.5 or less	f	(16)
NO_3^-	Laminaria	1.4×10^4 or less	f	(16)
PO_4^{3-}	Laminaria	3×10^5 or less	f	(16)
I^-	Laminaria	6×10^4	d	(9)

classified the mineral elements into 3 groups on the basis of their accumulation coefficient (Ka) in marine algae:

(1) elements with $Ka < 1 \times 10^2$ comprised Na, Br, Sr, Se.

(2) elements with $Ka > 1 \times 10^2$ but $< 1 \times 10^4$ comprised Rb, Cs, Ba, Sc, Co, Sb, Ag, Th, U.

(3) elements with Ka > 1 x 10^4 included Fe, Cr, Zn, Zr, Hf, As, Hg, Au, La, Nd, Yb.

Such a wide range of concentration factors and the fact that anions as well as cations may be accumulated or excluded by algal tissues (Table I) seems to indicate that different processes of uptake may be involved for different ions (for a detailed review see 45).

Ion competition and effect of external concentration on uptake

From the classical experiments of Osterhout on electrical resistance of _Laminaria_ tissues under various external ionic conditions (46), it is well known that a salt competition occurs with respect to cell membrane permeability, e.g., Ca^{++} and Na^+ are antagonistic to each other, so are Mg^{++} and K^+. Recent studies on _Laminaria_, _Fucus_, and _Macrocystis_ have shown that, contrary to the situation in phytoplankton, the presence of NH_4^+ does not inhibit NO_3^- uptake by these algae, and that both the ions are used simultaneously (29,48,49). However, in some other marine macrophytes, e.g., _Codium_, NH_4^+ does affect NO_3^- uptake (47). In addition, the uptake of NO_3^- and NH_4^+ by _Fucus_ and _Macrocystis_ is strongly affected by the external concentration of these ions (48,49). At relatively lower concentrations, the relationship between nutrient uptake and nutrient concentration in the medium is hyperbolic, and seems to follow Michaelis-Menten kinetics. However, there were deviations from simple hyperbolic relations at relatively high external concentrations of NO_3^- and NH_4^+, and the authors suggested that more than one process may be involved, for instance in the uptake of NH_4^+ by _Macrocystis_ (49), as is the case in phytoplankton (50).

Effect of light and darkness on uptake

Several papers have been published on the effect of light on ion exchange in large algae (Table II). It is well known that illumination has an immediate and a strong positive effect on potassium uptake, and to a lesser extent, on sodium extrusion (51,52). The effect of dark-light has

Table II: Effect of illumination on ion exchange in large algae. (+) positive effect, (-) no marked effect.

Ion	Algae	Effect	Reference
K^+, Na^+	Ulva lactuca	+	(51,52)
K^+, Na^+	Palmeria palmata	+	(53)
NH_4^+	Fucus spiralis	+	(48)
$^{134}Cs^{2+}$	Palmeria palmata	+	(54)
Ca^{2+}	Fucus vesiculosus	-	(55)
Zn^{2+}	Laminaria digitata	-	(39)
$^{131}I^-$	Ulva lactuca	+	(56)
NO_3^-	Enteromorpha sp.	+	(27)
NO_3^-	Laminaria longicruris	+	(29)
NO_3^-	Fucus spiralis	-	(48)

been interpreted in terms of a supply of K^+ carriers during photosynthesis in the light and their removal by respiration in the dark (53). Both K^+ uptake and Na^+ extrusion are considered to be regulated either by two active but separate mechanisms (52) or by Na^+ extrusion being dependent upon K^+ accumulation (57). In any case, such mechanisms require energy and hence can be related to metabolism.

Effect of metabolic inhibitors on uptake

Various metabolic inhibitors, such as iodoacetate, phenyl urethane dinitro-o-cresol, arsenate, potassium cyanide, and 2,4-dinitrophenol, have been reported to affect K^+ and Na^+ exchange in Ulva lactuca (51) and Hormosira banksii (58), Rb^+ uptake in Fucus serratus and Laminaria digitata (59), and SO_4^{2-} uptake in Fucus serratus (60).

Effect of temperature on uptake

As is true for light, the effect of temperature on ion uptake by large algae is ion specific and dependent on the species concerned (Table III). Thus, a strong negative effect in going from 15 C to 5 and 0 C has been

Table III: Effect of temperature on ion exchange in large algae. (+) positive effect, (-) no marked effect.

Ion	Algae	Effect	Reference
NO_3^-	Enteromorpha sp.	+	(27)
NO_3^-	Laminaria longicruris	+	(29)
NO_3^-	Fucus spiralis	-	(48)
$H_2PO_4^-$	Laminaria digitata	+	(23)
Zn^{2+}	Fucus, Enteromorpha	+	(61)
NH_4^+	Fucus vesiculosus	-	(55)
Ca^{2+}	Laminaria digitata	-	(23)

reported for NO_3^- uptake by Laminaria longicruris (29), in contrast to the results obtained with the same ion in Fucus spiralis (48). By the use of ^{32}P as a tracer, it has been shown that phosphate absorption in Laminaria digitata is much slower at 0 C than at 12 C, and that no leakage occurs for hours when the tissues incubated at 12 C are subsequently immersed at 0 C in sea water free of tracer (Fig. 1). Similar experiments carried out by the use of ^{45}Ca showed no significant difference in Ca^{++} uptake in L. digitata at 12 C and 0 C, which suggests that, contrary to the uptake of phosphorus, calcium uptake is not directly related to cellular metabolism. Moreover, when the ^{45}Ca labelled tissue was immersed in sea water free of tracer, a large proportion of the radio-calcium, 80% in the first 4 hours, was rapidly released into the medium, suggesting that most of it was sequestered in the intercellular free-space (Fig. 2; see also reference 23).

Mechanisms of uptake by brown algae

From the available data, two kinds of mechanisms seem to be involved in ion uptake by large brown algae. One is light and temperature dependent, requires energy and is inhibited by metabolic inhibitors; the other seems to be governed by physical processes, and is apparently independent of metabolism.

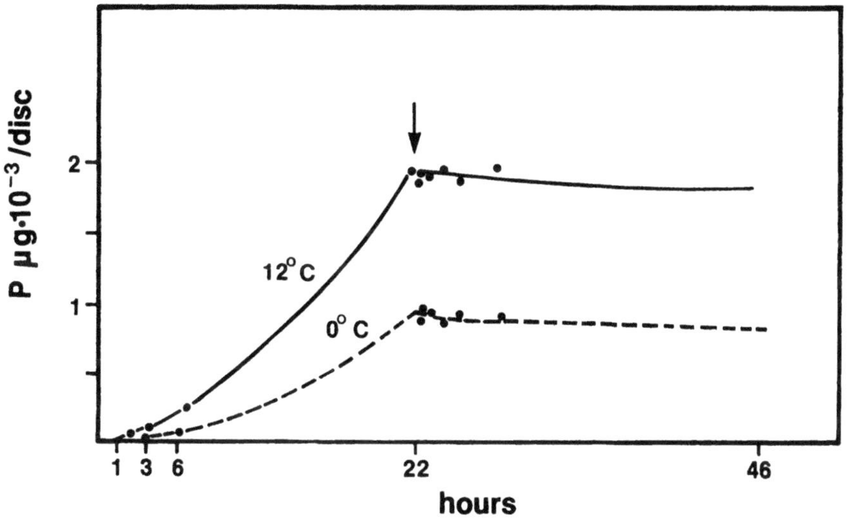

Fig. 1: Time course of absorption and exsorption of phosphorus by discs punched from the middle of the lamina of Laminaria digitata. Absorption in filtered sea water at 0 and 12 C. Exsorption in filtered sea water at 0 C (arrow). Amount of P taken up or retained by the tissues is expressed as micrograms per 12 mm diameter disc. (From Floc'h and Penot [23])

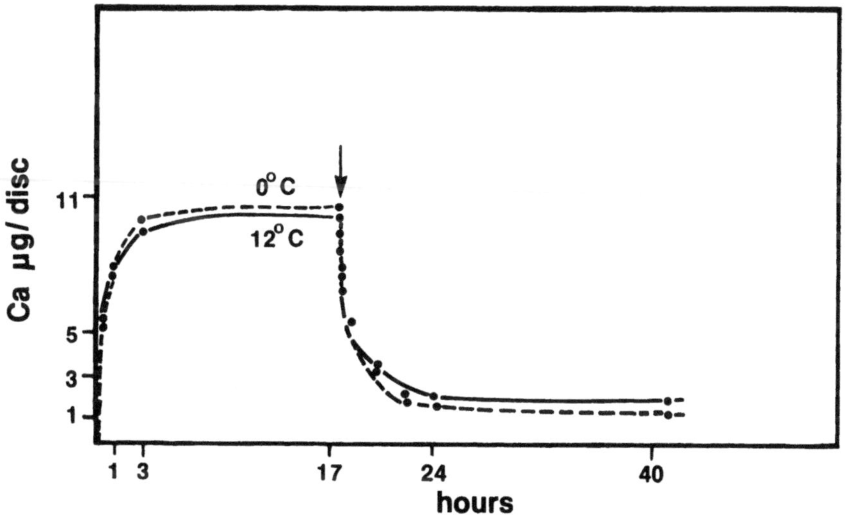

Fig. 2: Time course of absorption and exsorption of calcium by Laminaria digitata. Same conditions as in Fig. 1. (From Floc'h and Penot [23])

It is well known that two of the most characteristic features of brown algae are:

(1) the great amount of negatively charged polysaccharides in their cell walls;

(2) the occurence of various types of vesicles in the protoplasm, vacuoles or physodes, whose composition and role are not well known.

In this connection, some of the earlier work on zinc uptake in brown algae is relevant. Gutknecht (62) studied zinc uptake in several marine algae, including _Fucus_, and concluded that the main uptake mechanism for zinc was a simple ion-exchange reaction with the acidic polysaccharides of the cell wall. The passive metal accumulation in brown algae, as suggested by some authors (38,63), would be consistent with such a hypothesis. However, although there was no evidence that Zn^{++} uptake in _Laminaria_ could be regulated by metabolism, Bryan (39) suggested that the intercellular polysaccharides were not much involved in Zn^{++} binding. Skipnes _et al._ (64) provided further support for this idea: while strontium accumulation seemed to be an ion-exchange process involving the negatively charged wall polysaccharides, only a fraction of the gradual, long-term uptake of Zn^{++} by _Ascophyllum_ seemed to be due to an ion-exchange process. Skipnes _et al._ considered that the intercellular charged polysaccharides acted as ion buffers, allowing Zn^{++} uptake into the cell at a constant rate independent of tidal height. They suggested that Zn^{++} was bound to substances contained in membrane-bound vesicles, probably vacuoles. Interestingly, a recent study (65) on metal chelation by phloroglucinols from _A. nodosum_ and _Fucus vesiculosus_ agrees with this hypothesis, and the authors suggest that physodes could be one of the possible sites for such natural chelators of metals. It must be pointed out, however, that accumulation of Zn^{++} in physodes would not occur without a prior entry of Zn^{++} into the cell which may be dependent upon metabolism. The relationship between the Zn^{++} uptake by _F. virsoides_ and the temperature-dependent metabolic activity of the alga, as suggested by Munda (61), is consistent with this idea.

As regards sulfated polysaccharides of brown algae, it has been suggested that fucoidan plays an important part in ^{35}S uptake by Fucales (60,66), but the actual role of the cell walls of brown algae as _in situ_ ion-exchangers is not known.

Preferential distribution of ions and their uptake

A nonhomogeneous distribution of phosphate (13), and potassium and iodine accumulation (14) has been reported previously in different parts of the Laminaria thallus. A more detailed study showed a preferential distribution of several mineral elements along the thallus of various members of the Laminariales, which was specific to the ions and related to the age of the tissues and their function, e.g., blade, holdfast, meristem, etc. (Figs. 3,4; see also references 17,18). A similar distribution of phosphorus occurs along the thallus of Fucus vesiculosus (16). Moreover, analysis of P, K, Na, Ca and Mg, performed from December to April, in blades of growing L. hyperborea, showed that the distribution pattern of ions between the older and younger parts of the lamina changes during the growth of the alga and that this variation is ion specific (24).

Since uptake of mineral elements may continue throughout the life of an algal tissue (38,39), this longer duration of uptake may explain the larger accumulation of ions, such as K^+ and Ca^{++} in the older tissues of Laminaria hyperborea (24). By contrast, a lower concentration of ions in those same tissues may be explained by reduced uptake of these ions by older fronds, as seems to occur for NH_4^+ and NO_3^- in Fucus spiralis (48), or by their greater extrusion. However, a comparison of ^{32}P uptake by the younger and older laminae of the same thallus of L. hyperborea did not show any significant difference in the tissue's capacity for uptake or extrusion (Fig. 5). A similar result was reported for blade discs of Macrocystis (67). Therefore, it appears that at least the heterogeneous distribution of phosphorus along the thallus of the Fucales and Laminariales cannot be explained by a differential exchange ability of the different regions and tissues. As a result, as postulated by Wille (13), a translocation of minerals from one part of the thallus to another must be taken into account.

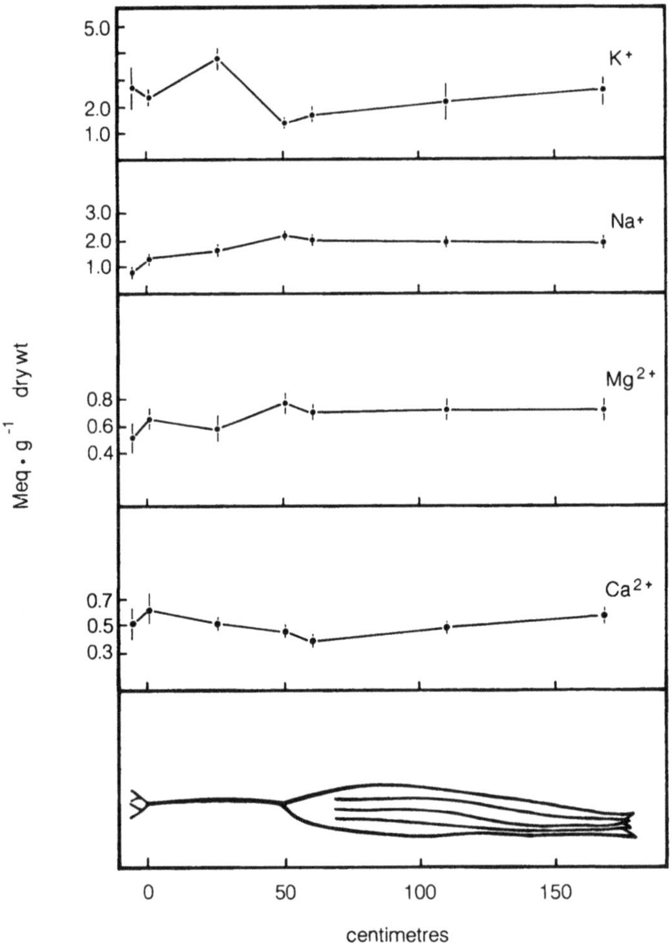

Fig. 3: Distribution of potassium, sodium, magnesium and calcium along the thallus of Laminaria digitata. Contents expressed in milliequivalents per gram of dry weight. Means of four series of measurements and standard deviations are presented. Samples collected in September. Measurements by flame spectrometry (K,Na) and by complexometry (Mg,Ca).

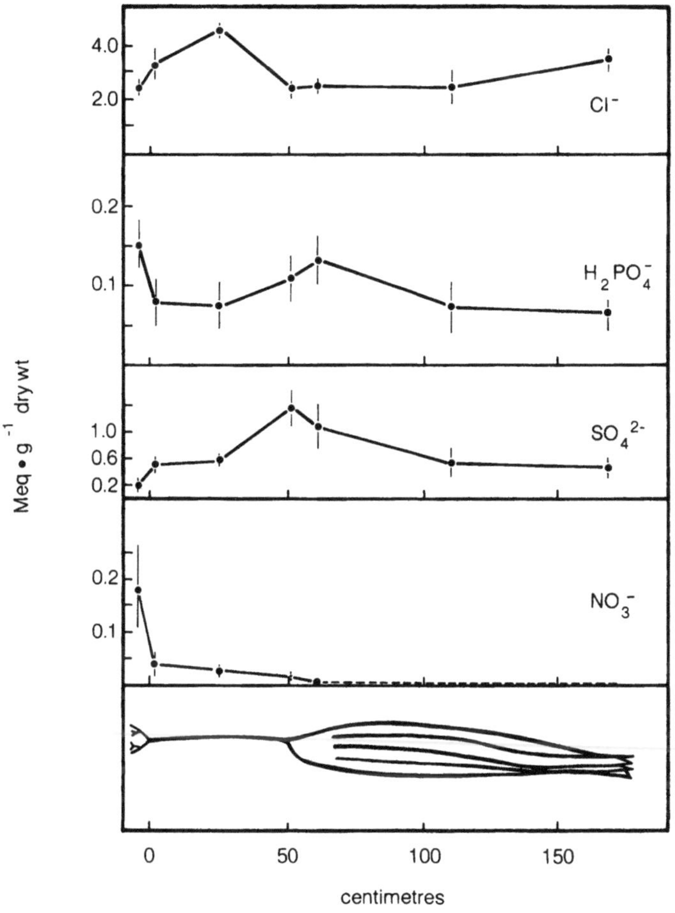

Fig. 4: Distribution of chloride, phosphate, sulfate and nitrate along the thallus of Laminaria digitata. Contents expressed in milliequivalents per gram of dry weight. Means of four series of measurements and standard deviations are shown. Samples collected in Semptember. Measurements made by colorimetric methods.

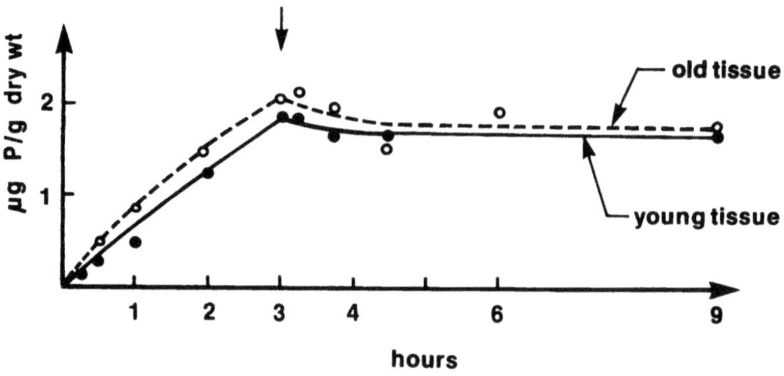

Fig. 5: Comparative phosphorus uptake and release by an old and a very young lamina of the same thallus of Laminaria hyperborea in January. Uptake measured by the use of ^{32}P in carrier-free sea water: exsorption (arrow) in non-labelled sea water. T = 12 C. Results expressed in μg of P taken up per g dry weight of algal tissue.

Long Distance Transport of Ions

Although translocation of inorganic substances in algae has been doubtful for a long time (68,69), a long distance transport of mineral elements has been demonstrated directly in the last ten years in many of the Laminariales and Fucales (Table IV, see also references 15-24,67).

Ion specificity

By the use of various tracers (^{32}P, ^{86}Rb, ^{35}S, ^{99}Mo, ^{45}Ca, ^{36}Cl), it has been shown that mineral elements do not move identically in the thallus of Laminaria digitata: phosphorus, sulfur and rubidium show directed long distance transport, whereas chloride, molybdenum and calcium do not seem to move (21).

The need for phosphorus in growing or meristematic regions is well known, so is the fact that these regions are strong sinks. This transport will be reviewed in detail further on.

Table IV: Brown algae in which directed long distance transport of mineral elements has been reported.

Algae	Translocated Isotope	Reference
FUCALES		
Fucaceae		
Fucus vesiculosus	^{32}P, ^{86}Rb, ^{35}S	(19,20)
Fucus spiralis	^{32}P	(Fig. 6)
Fucus serratus	^{32}P	(16)
Fucus ceranoides	^{32}P	(a)
Cystoseiraceae		
Cystoseira baccata	^{32}P	(Fig. 6)
LAMINARIALES		
Laminariaceae		
Laminaria digitata	^{32}P, ^{86}Rb, ^{35}S	(18,21,23)
Laminaria hyperborea	^{32}P, ^{86}Rb	(19,24)
Laminaria setchelli	^{32}P	(Fig. 7)
Laminaria ochroleuca	^{32}P	(16)
Laminaria saccharina	^{32}P	(16)
Chordaceae		
Chorda filum	^{32}P	(Fig. 6)
Phyllariaceae		
Saccorhiza polyschides	^{32}P	(Fig. 6)
Alariaceae		
Alaria esculenta	^{32}P, ^{86}Rb	(19)
Alaria marginata	^{32}P	(Fig. 7)
Egregia menziesii	^{32}P	(Fig. 7)
Lessoniaceae		
Macrocystis integrifolia	^{32}P	(67)
Nereocystis luetkeana	^{32}P	(Fig. 7)

(a) = unpublished data

Sulfur translocation occurs in Laminaria digitata towards the meristem (21), where chemical analysis also shows an accumulation of the cold element (Fig. 4). ^{35}S was also shown to move towards the apical meristem in Fucus vesiculosus (unpublished results). There is no detailed information in the literature on the distribution of the sulfated polysaccharide fucoidan in the different regions of the thallus of brown algae. Therefore, it is not possible to relate the long distance transport of ^{35}S to any region that may accumulate this polysaccharide. Sulfur translocation may also be related to protein synthesis which is usually high in meristems.

Translocation of ^{86}Rb has been shown to occur towards the growing young tissues of Laminaria and Fucus. Whether ^{86}Rb can be considered as a tracer for K^+ in algae, as it is in higher plants (70), is a moot point. West and Pitman (71) compared ^{86}Rb and ^{42}K uptake by Ulva and concluded that the two elements had different rates of uptake. A correspondence between ^{86}Rb translocation and K^+ accumulation was evident only for a short period (December to February) in the very young meristematic tissue of the new lamina of L. hyperborea (21), and, therefore ^{86}Rb could be considered as a tracer for K^+ in a limited way. There is very little information, to date, on K^+ translocation in algae, although this ion is present in a high concentration in sieve-tube exudate of Macrocystis (9.2 to 12.6 mg ml^{-1}, see references 67,72), and in medulla exudate of Nereocystis (7.7 mg ml^{-1}, see reference 73).

The immobility of some of the mineral elements when fed to algae may result either from a linkage to structural components, or from an exclusion from the algae. The latter phenomenon would explain the immobility of ^{36}Cl when injected into Laminaria tissue, since there is less chloride in the alga than in the sea water (concentration factor = 0.5 in the lamina, see reference 21).

Molybdenum is a cofactor of nitrate reductase (74). Although NO_3^- distribution is not homogeneous in the thallus of Laminaria (Fig. 4; see also reference 75), little is known about the relative activity of nitrate reductase in different parts of the thallus (76,77). Since ^{99}Mo did not move, it is possible that the internal concentrations of this element are sufficient for the enzyme activity in different parts of the thallus.

Since calcium uptake does not seem to be directly controlled by cellular metabolism (Fig. 2), the immobility of ^{45}Ca may be related to calcium binding to acidic wall polysaccharides in brown algae (21,24), especially in the older tissues of Laminaria which are known to be rich in guluronic acid fractions with a strong affinity for calcium (78).

Thus, it appears that in brown algae, both uptake and translocation of mineral elements are ion specific. Since most of the research on long distance transport of mineral elements in algae has been carried out with ^{32}P, the following review will deal with the characteristics of its transport.

Transport of ^{32}P

Directionality, "source to sink" relationship. In all studies dealing with the Fucales and Laminariales, ^{32}P translocation occurs from the older tissues toward the younger growing regions of the thallus. Whatever the fed region in the laminae of the Laminariales, ^{32}P moved in a basipetal direction toward the meristematic transition zone, the holdfast, and the sporophylls, never toward the older tissues (Figs. 6,7). In Chorda filum ^{32}P transport is acropetal (Fig. 6) and, interestingly, the meristem is subterminal in that alga (79); since ^{32}P is not translocated above the meristem, this apparent exception in the Laminariales is in good agreement with the general rule that transport is toward the young growing tissues.

A comparison of ^{32}P accumulation along the thalli of the Laminariales (Fig. 7, see also reference 20) and the cold phosphorus accumulations as determined by chemical analyses (Fig. 4), indicates that translocation is an important process which links the initial site of phosphate entry to the final site of accumulation or consumption. In the Laminariales the main sinks or sites of accumulation seem to be the transition zone between the lamina and the stipe, the meristoderm of the stipe, the base of the stipe, the younger haptera of the holdfast, and the growing sporophylls (as in Alaria, Fig. 7).

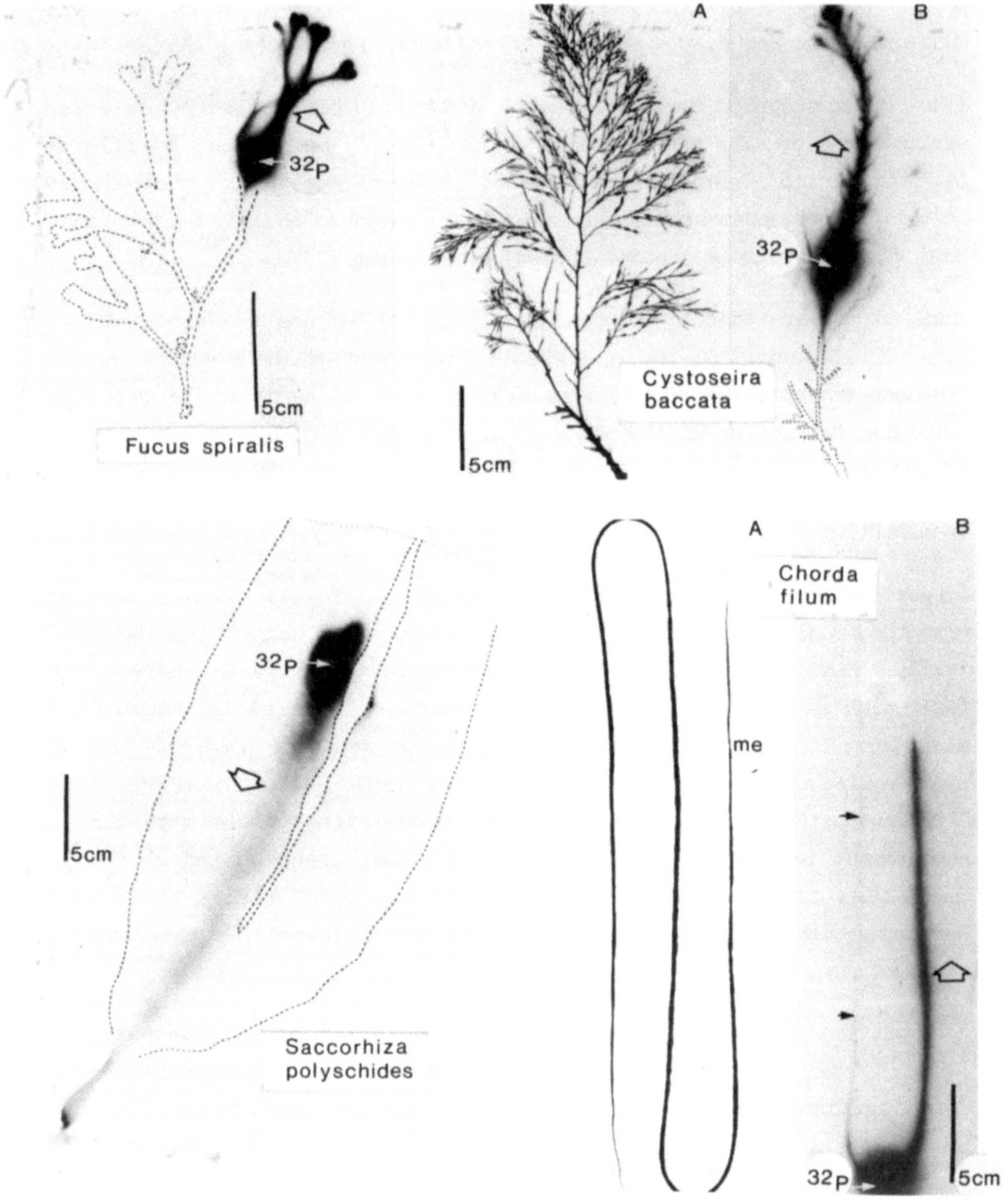

Fig. 6: Autoradiographs of brown algae labelled with ^{32}P. Uptake: 3 h (white arrows); migration: 15 days in situ (large arrows). For Fucus and Sacchorhiza, the outlines of thalli are delineated by dotted lines. For Cystoseira and Chorda, (A) represents the whole thallus, (B) shows the autoradiograph after translocation of the radioisotope. For Chorda, me: meristem; the small black arrows indicate an occasional slight generalized labelling which could be an artifact.

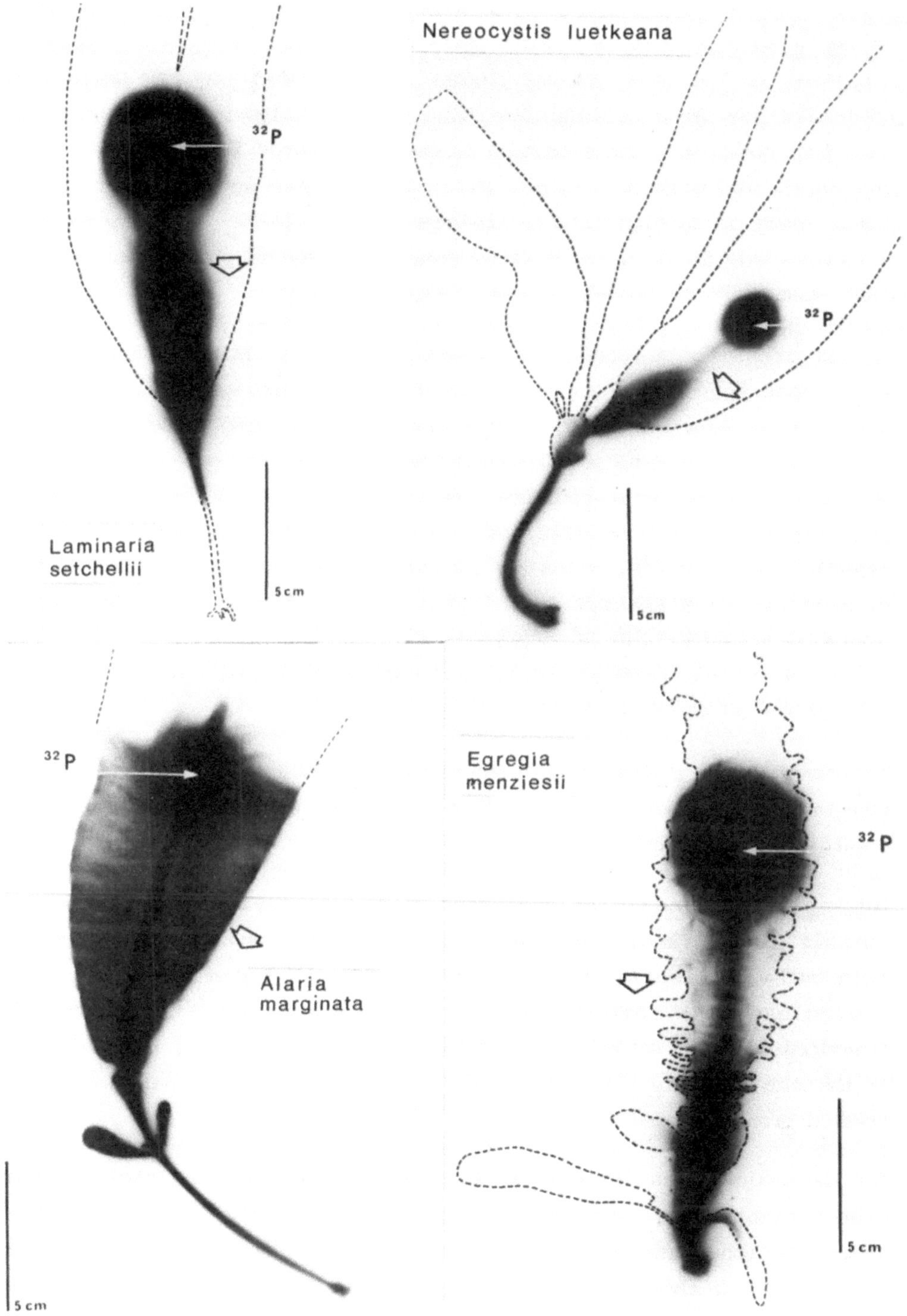

Fig. 7: Autoradiographs of brown algae labelled with ^{32}P. Uptake: 3 h (white arrows); migration: 24 h <u>in situ</u> (large arrows).

In Laminaria hyperborea the older parts of the blade serve as a source of phosphate for the growth of the younger parts, in much the same way as has been shown for carbon assimilates (80). Specifically, for phosphorus, it has been shown that concentration of cold phosphorus increases rapidly in the young, fast-growing lamina, while it declines in the older lamina (24). There is no significant difference in ^{32}P uptake by young and old tissues, and, finally, there is no greater leakage of phosphorus from the older than from the younger tissues (Fig. 5).

Translocation of ^{32}P occurs in the Fucales as well. In Fucus vesiculosus the tracer moved to the apical meristem and to the basal disc (Fig. 6), where the amounts of cold phosphorus also had been found to be the highest (17,19). In these algae, the apical meristem and the basal disc seem to be the main sinks for phosphorus. Removal of the apical meristem in F. vesiculosus and the holdfast of Laminaria digitata slows down the translocation of ^{32}P, but this is followed by wound healing which stimulates the migration of phosphate (15). Moreover, an exceptional acropetal translocation of ^{32}P in L. digitata, toward a region of the lamina which had been experimentally wounded, showed that regenerating tissues act as strong sinks.

The above review indicates that phosphate translocation in brown algae is regulated by a "source to sink" relationship, similar to the pattern of mineral translocation in higher plants.

Pathway of transport. As suggested by the intensity of labelling in autohistoradiograms, the midrib of many of the Fucales seems to be the main pathway of mineral transport (19). If the midrib of Fucus vesiculosus was removed, leaving only the wings, there was no translocation of ^{32}P toward the apex of the treated branch; by contrast, if the midrib was left intact and the wings removed, ^{32}P moved through the treated branch as well as through the control.

For the Laminariales, an experiment, based on the classical "stripping" in higher plants (81), was carried out to define the pathway of ^{32}P translocation. The cortical tissue in Laminaria digitata stipe was separated from the medulla by two impermeable sheets and the migration of the radioisotope from the fed lamina through the different tissues of the stipe was followed. The results showed clearly that the medulla was the pathway of the basipetal long distance transport of ^{32}P, and that a secondary lateral transport occurred from the medulla to the meristoderm which was stopped by the impermeable sheets (Fig. 8). Since the meristoderm of the stipe is a well known meristematic tissue in the Laminariales, it is not surprising that it acts as a "sink" for mineral translocation.

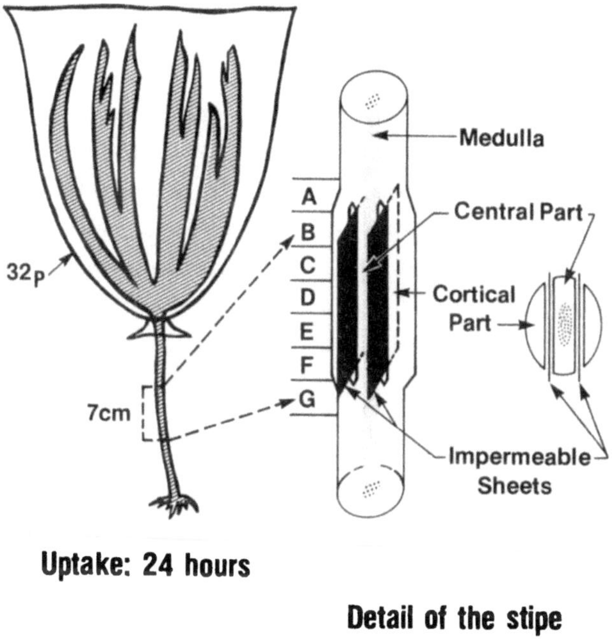

samples	cortex	centre	cortex
A	520	1964	730
B	436	1337	245
C	7	1393	13
D	6	1280	7
E	8	1218	8
F	38	1141	22
G	570	1255	200

32p dpm/100 sec/100 mg fresh wt
Translocation: 6 days in situ

Results

Fig. 8: Distribution of ^{32}P in different parts of the stipe of Laminaria digitata. Longitudinal incisions separated the medulla from the cortex and impermeable sheets were inserted between the two. (Adapted from [19])

We do not know which part of the tissue or cells in the medulla of the Laminariales are involved in ion translocation, but since ^{32}P has been found in the sieve tube sap of Macrocystis and shows the same velocity and directionality of transport as ^{14}C (67), it most probably moves through the sieve elements (for ^{14}C translocation in the sieve elements of the Laminariales, see Schmitz, this volume). Further research in microlocalization of ^{32}P is necessary to confirm this assumption.

Velocity, rate, seasonality. Since phosphate translocation occurs against a concentration gradient in brown algae, this transport is not a case of mere diffusion, as thought earlier (69), but is related to algal metabolism. Moreover, the velocities of transport, as measured in several of the Laminariales (2 to 45 cm h^{-1}, see references 19, 20,67), are a good deal faster than those measured in killed tissues (0.06 cm h^{-1}, reference 20), or those reported earlier by Parker (69) for Fucus vesiculosus and Laminaria agardhii ($v < 0.08$ cm h^{-1}). In addition to the heavy labelling due to directional transport, some autoradiograms show a light, homogeneous labelling of the thallus (Chorda filum, Fig. 6). Such a labelling has been reported to go across the barrier between the host alga, Ascophyllum nodosum, and its epiphytes, in both directions, irrespective of the radioisotopes used (82), and interpreted as a very fast transport -- up to several meters h^{-1} (83). Such a high velocity of transport has been reported in land plants as well, though some authors consider it an artifact (84).

Measurement of the amounts of translocated ^{32}P in young thalli (age $\leq$ 1 year) of Laminaria digitata from the source (middle of the lamina) to the sinks (base of the lamina, stipe, holdfast), showed that less than 1% of the total radiophosphorus taken up was translocated in 24 h (16). This rate, which was obtained during December, is rather weak and probably larger amounts are translocated during the period of active growth. Since phosphate absorption presumably occurs over the whole lamina, the total amount of phosphate supplied to the growing tissues by the source through translocation does not seem to be so insignificant.

The relationship between source and sink varies during the year and is closely related to the growth pattern of the alga. Thus, in Laminaria digitata, ^{32}P translocation starts in the fall when the growth of the alga starts, and decreases sharply or stops in summer, when the growth slows down. Moreover, it has been shown that different parts in the growing region of the lamina of L. digitata may accumulate different amounts of ^{32}P from long distance transport and thus may have different "sink" capacities (20).

Effect of cycloheximide on long distance transport of ^{32}P. Cycloheximide, an inhibitor of protein synthesis, was applied to three regions of Laminaria digitata, the old part of the lamina (traditional "source"), the young meristematic region (traditional "sink") and the intermediate region ("pathway") and its effect on ^{32}P transport was tested after supplying the radioisotope to the old part (22). The results showed that when cycloheximide was applied to the base of the lamina, there was no ^{32}P translocation into this region. In other words, this part no longer acted as a sink. When cycloheximide was applied between the base of the lamina and the ^{32}P-fed region, the radioisotope did not move past the cycloheximide-treated region. Finally, when cycloheximide was applied to the same region as ^{32}P, either as a pretreatment or simultaneously, the radioisotope was retained at the source. Although the effects of cycloheximide on photosynthesizing tissues of brown algae are not well known, these results may indicate a dependence of ^{32}P loading as well as its long distance transport on continued protein turnover. The lack of translocation of ^{32}P to the traditional sink area is more readily explained by an inhibition of protein synthesis and a consequent inhibition of meristematic and growth activity.

Possible forms of phosphate transport and accumulation. Floc'h and Penot (23) studied the patterns of ^{32}P labelling in phosphorylated compounds in regions of uptake, translocation and accumulation in Laminaria digitata. In the uptake region (lamina), 3 h after the beginning of incubation, the bulk of radioactivity was incorporated into organic compounds, hexose monophosphates being most heavily labelled. This result seems to confirm the metabolic dependence of phosphorus uptake (Fig. 1). Subsequently there was a decrease in radioactivity in the organic phosphate fraction and a simultaneous ^{32}P accumulation in the inorganic phosphate pool: after a period of 13 days, 65-70% of the ^{32}P taken up was found in the inorganic fraction. In the medulla of the stipe (the conducting zone), the pattern of ^{32}P phosphorylated compounds was similar to that in the uptake region in the first few hours, i.e. hexose monophosphates were the most labelled intermediates, and this pattern was maintained throughout the 13 day experiment. In the accumulating tissues (the sink), the pattern of ^{32}P- phosphorylated compounds was at first similar to that in the uptake and conducting zones, but subsequently there was an increase of labelling simultaneously in the inorganic fraction and in the insoluble organic fraction of phosphate. Floc'h and Penot (23) suggested that, since the ratio of P_i/P ester remained constant in the conducting zone, while it increased considerably in the region of uptake as well as in the sink, inorganic phosphate could be one of the possible translocating forms. The slow transfer of ^{32}P into the inorganic pool in the uptake region, as shown by Floc'h and Penot (23) in L. digitata, could explain why no ^{32}P translocation was found in Alaria marginata 5 h after feeding, while ^{14}C, applied simultaneously, was translocated during that time (85).

More recently, Schmitz and Srivastava (67) reported on the pattern of ^{32}P-phosphorylated intermediates in the fed region (blade) as well as in the sieve tube sap of Macrocystis integrifolia. Their results are in good agreement with those obtained for Laminaria in terms of both the high amount of ^{32}P found in the inorganic form in the fed region and the high levels of labelling of hexose monophosphates in the sieve tube sap. Moreover, for the first time, these authors reported that double labelled ^{32}P and ^{14}C hexose monophosphates occured in the sieve tube sap of M. integrifolia along with ^{14}C labelled mannitol and amino acids. These authors suggested that phosphorus was probably translocated in an organically bound form, but this does not exclude the possibility that phosphorylation may still occur in the sieve tube sap (86).

The available data on phosphorus uptake, translocation and accumulation in brown algae show several parallels with the available data on phosphate uptake and translocation in higher plants. The slow, but continuous ^{32}P transfer from the soluble organic fraction to the inorganic fraction after uptake in Laminaria digitata is in good agreement with the results obtained by Bieleski and Laties in potato tuber tissue (87). These authors found that phosphorus went first to a metabolic pool and then to a nonmetabolic pool, with the time of equilibration being fairly long. In addition, Bieleski (88) suggested that, since inorganic phosphorus is the first phosphate form to enter the phloem, it seems to be the translocating form in land plants. Many other similarities have been found between mineral translocation in algae and in higher plants (24,89).

An attempt has been made to explain the mechanism of long distance transport in large algae by considering the mass-flow theory of Münch (see Schmitz, this volume). However, other mechanisms are possible. "Activated diffusion," which postulates that "the tissues receiving from and supplying the pathway play an active part in the interchange of solutes" (90), either by means of a push at the source or a consumption at the sink, could also be applied to the algae as well, especially since the source as well as the sink tissues of the Fucales and the Laminariales have been shown to play an active part in translocation. Moreover, some aspects of the electro-osmotic theory (91) may find support, as a relay, in ion translocation in marine macrophytes since the cell walls of these plants are peculiarly rich in negatively charged molecules and since enzymatic activity occurs in the sieve tube sap which is rich in K^+, features which are essential to this theory.

Finally, the demonstration that a long distance transport of mineral elements occurs in algae (15-24,67) opens up many new questions. For instance, in what chemical form(s) is phosphate translocated or in what form(s) are various other mineral elements loaded, unloaded, and transported, and what is the site of such transport and accumulation are questions for which we still have no answers.

Acknowledgements

I am very grateful to Professor L.M. Srivastava, Simon Fraser University, for providing laboratory facilities during my tenure as a Visiting Research Associate in the Department of Biological Sciences, Simon Fraser University, Burnaby, B.C., Canada, and for critically discussing the manuscript, and helping to write it in English.

References

1. Jensen, A.: Proc. Int. Seaweed Symp. 9, 17-34 (1978).
2. Hendrick, J.: J. Soc. Chem. Ind. 35, 565-574 (1916).
3. Vincent, V.: Bull. Sta. Agr. Fin., Quimper, 206 pp (1924).
4. Black, W.A.P., Mitchell, R.L.: J. mar. biol. Ass. UK 30, 575-584 (1952).
5. Vinogradov, A.P.: Mem. Conn. Yale Univ. Press 11, 647 pp. (1953).
6. Ishibashi, M., Yamamoto, T.: Rec. oceanogr. Works, Jap. 5, 55-62 (1960).
7. Lunde, G.: J. Sci. Food Agric. 21, 416-418 (1970).
8. Saenko, G.N., Koryakova, M.D., Makienko, V.E., Dobrosmyslova, I.G.: Mar. Biol. 34, 169-176 (1976).
9. Gordon-Young, E., Langille, W.M.: Can. J. Bot. 36, 301-310 (1958).
10. Larsen, B., Jensen, A.: Rep. Norw. Inst. Seaweed Res. 15, 22 pp. (1957).
11. Strauss, R.: Int. Revue ges. Hydrobiol. 52, 465-486 (1967).
12. Jensen, A.: In Synopsis of the Biological Data on Laminaria hyperborea (J. Kain, ed.), FAO Fish. Synop. 87, 65 pp. 1971
13. Wille, N.: Bot. Untersuch. Berlin, 321-340 (1899).
14. Rinck, E., Brouardel, J.: C.R. Acad. Sci. Paris 229, 1167-1168 (1949).
15. Floc'h, J.Y.: Phys. Vég. 14, 767-777 (1976).
16. Floc'h, J.Y.: Etude du transport a longue distance des éléments minéraux dans le thalle des algues brunes. Thèse Doct. d'Etat, Brest 1979
17. Floc'h, J.Y., Penot, M.: C.R. Acad. Sci., Paris 271, 288-291 (1970).
18. Floc'h, J.Y., Penot, M.: C.R. Acad. Sci., Paris 273, 1100-1103 (1971).
19. Floc'h, J.Y., Penot, M.: Phys. Vég. 10, 677-686 (1972).
20. Floc'h, J.Y., Penot, M.: Proc. Int. Seaweed Symp. 8, (in press).
21. Floc'h, J.Y., Penot, M.: C.R. Acad. Sci., Paris 282, 989-992 (1976).
22. Floc'h, J.Y., Penot. M.: Bot. Mar. 21, 5-11 (1978).
23. Floc'h, J.Y., Penot, M.: Planta 143, 101-107 (1978).
24. Floc'h, J.Y., Penot, M.: Z. Pflanzenphysiol. 96, 377-385 (1980).
25. North, W.J., Wheeler, P.A.: Proc. Int. Seaweed Symp. 9, 263-271 (1978).
26. Ryther, J.H., DeBoer, J.A., Lapointe, B.E.: Proc. Int. Seaweed Symp.

9, 1-16 (1978).

27. Harlin, M.M.: Aquaculture 15, 373-376 (1978).

28. DeBoer, J.A.: Proc. Int. Seaweed Symp. 9, 263-271 (1978).

29. Harlin, M.M., Craigie, J.S.: J. Phycol. 14, 464-467 (1978).

30. Kain, J.M.: Estuarine Coastal Mar. Sci. 7, 531-553 (1978).

31. Wong, M.H., Lau, K.K.: Chemosphere 8, 217-224 (1979).

32. Phillips, J.H.: Environ. Poll. 18, 31-43 (1977).

33. Takashima, Y., Momoshima, N., Nakayama, Y., Yamashita, H., Nakashima, H.: Radiochem. Radioanal. Lett. 36, 61-67 (1978).

34. Shiber, J.G., Shatila, T.: Hydrobiology 63, 105-112 (1979).

35. Taka, Y., Stara, J.F.: In Marine Algae in Pharmaceutical Science (H.A. Hoppe, T. Levring, Y. Tanaka, eds.) Walter de Gruyter, Berlin, pp. 525-543 1979

36. Poehlmann, W., Rietz, K.: Muench. Beitr. Abwasser-, Fish.-Flussbiol. 30, 121-127 (1978).

37. Watanabe, T., Hirayama, T., Takahashi, T., Kokubo, T., Ikeda, M.: Toxicology 14, 1-22 (1979).

38. Morris, A.W., Bale, A.J.: Estuarine Coastal Mar. Sci. 3, 153-163 (1975).

39. Bryan, G.W.: J. mar. biol. Ass. UK 49, 225-243 (1969).

40. Mauchline, J., Templeton, W.L.: J. Cons. perm. int. Explor. Mer. 30, 161-170 (1966).

41. Whyte, J.N.C., Englar, J.R.: Bot. Mar. 23, 13-17 (1980).

42. Saenko, G.N., Karyakin, A.V., Koryakova, M.D., Makienko, V.F., Dobrosmylova, I.G.: In Kzaim. Vod. Zhiv. Veshch., Tr. Meshd. Simp. (E.V. Krasnov, ed.), Izd. Nauka, Moscow, pp. 164-169 1975

43. Seeliger, U., Edwards, P.: Mar. Poll. Bull. 8, 16-19 (1977).

44. Vaganov, P.A., Kulikov, V.D., Shtangeeva, I.V.: Geokhimiya 11, 1740-1745 (1978).

45. Gutknecht, J., Dainty, J.: Ann. Rev. Oceanogr. Biol. 6, 163-200 (1968).

46. Osterhout, W.J.V.: Injury, Recovery and Death in Relation to Conductivity and Permeability. J.B. Lippencott, Philadelphia 1922

47. Hanisak, D., Harlin, M.: J. Phycol. 14, 450-454 (1978).

48. Topinka, J.A.: J. Phycol. 14, 241-247 (1978).

49. Haines, K.C., Wheeler, P.A.: J. Phycol. 14, 319-324 (1978).

50. Falkowski, P.G., Rivkin, R.B.: J. Phycol. 12, 448-450 (1976).

51. Scott, G.T., Hayward, H.R.: J. Gen. Physiol. 36, 659-671 (1953).

52. Scott, G.T., Hayward, H.R.: J. Gen. Physiol. 37, 601-620 (1954).

53. MacRobbie, E.A.C., Dainty, J.: Physiol. Plant. 11, 782-801 (1958).

54. Scott, R.: Proc. 2nd Radioisotope Conference, Oxford. Medical and Physiological Applications. pp. 373-380 (1954).

55. Swift, E., Taylor, W.R.: Biol. Bull. 119, 342 (1960).

56. Roche, J., Andre, S.: C.R. Soc. Biol. 157, 1412-1416 (1963).

57. Eppley, R.W.: J. Gen. Physiol. 43, 29-38 (1959).

58. Berquist, P.L.: Physiol. Plant. 11, 760-770 (1958).

59. Penot, M., Videau, C.: Z. Pflanzenphysiol. 76, 285-293 (1975).

60. Coughlan, S.: J. Exp. Bot. 28, 1207-1215 (1977).

61. Munda, I.M.: Bot. Mar. 22, 149-152 (1979).

62. Gutknecht, J.: Limnol. Oceanogr. 10, 58-66 (1965).

63. Fuge, R., James, K.H.: Mar. Poll. Bull. 5, 9-12 (1974).

64. Skipnes, O., Roald, T., Haug, A.: Physiol. Plant. 34, 314-320 (1975).

65. Ragan, M.A., Smidsrød, O., Larsen, B.: Mar. Chem. 7, 265-271 (1979).

66. Lestang-Bremond, G., Quillet, M.: Phys. Vég. 14, 259-269 (1976).

67. Schmitz, K., Srivastava, L.M.: Plant Physiol. 63, 1003-1009 (1979).

68. Bodenberg, E.T.: Publ. Puget Sound Biol. Stn. 5, 253-256 (1928).

69. Parker, J.: Naturwissenschaften 43, 452 (1956).

70. Epstein, E., Hagen, C.E.: Plant Physiol. 27, 457-473 (1952).

71. West, K.R., Pitman, M.G.: Nature 214, 1262-1263 (1967).

72. Parker, B.C.: J. Phycol. 2, 38-41 (1966).

73. Nicholson, N.L., Briggs, W.R.: Am. J. Bot. 59, 97-106 (1972).

74. O'Kelley, J.C.: In Algal Physiology and Biochemistry (W.D.P. Stewart, ed.), Blackwell, Oxford, pp. 610-635 1974

75. Citharel, J.: Contribution à l'étude du métabolisme azoté des algues marines. Thèse Doct. d'Etat, Rennes, France 1971

76. Takagi, M., Murata, K.: Bull. Fac. Fish. Hokkaido Univ. 4, 306-309 (1954).

77. Takagi, M., Murata, K.: Bull. Fac. Fish. Hokkaido Univ. 6, 25-28 (1955).

78. Haug, A., Larsen, B., Baardseth, E.: Proc. Int. Seaweed Symp. 6, 443-451 (1969).

79. South, G.R., Burrows, E.M.: Br. Phycol. Bull. 3, 379-402 (1967).

80. Lüning, K., Schmitz, K., Willenbrink, J.: Mar. Biol. 23, 275-283 (1973).

81. Stout, P.R., Hoagland, D.R.: Am. J. Bot. 26, 320-324 (1939).

82. Penot, M.: Z. Pflanzenphysiol. 73, 125-131 (1974).

83. Penot, M., Penot, M.: Z. Pflanzenphysiol. 95, 265-273 (1979).

84. Hoddinott, J., Gorham, P.R.: Can. J. Bot. 54, 2415-2420 (1976).

85. Schmitz, K., Srivastava, L.M.: Can. J. Bot. 53, 861-876 (1975).

86. Schmitz, K., Srivastava, L.M.: Planta 116, 85-89 (1974).

87. Bieleski, R.L., Laties, G.G.: Plant Physiol. 38, 586-594 (1963).

88. Bieleski, R.L.: Plant Physiol. 44, 497-502 (1969).

89. Penot, M., Floc'h, J.Y., Penot, M.: Phycologia 15, 299-308 (1976).

90. Mason, T.G., Phillis, E.: Bot. Rev. 3, 47-71 (1937).

91. Spanner, D.C.: In Encyclopedia of Plant Physiology, New Series, Transport in Plants, I. Phloem Transport (M.H. Zimmermann, J.A. Milburn, eds.), Springer, Berlin, pp. 301-327 1975

DISCUSSION

WILLENBRINK: From which results do you conclude that it is active transport of phosphate? Of course, it could be active transport in an indirect manner.

FLOC'H: Since the velocities are higher than diffusion, and since the transport occurs against concentration, we have to suppose that metabolism is involved in that translocation. I think it could be related to the theory sometimes accepted in higher plants - the so-called activated diffusion theory. Because the transport is toward the younger parts of the plant, where the phosphate is consumed, we can suppose that this is one of the indirect effects of metabolism on the transport.

SRIVASTAVA: I was intrigued by the distribution of Pi and UDPG and hexose-phosphates and so on, in the three areas. Even after 4 days and 13 days the relative concentrations of these things in the conducting zone, not in the fed or accumulating zone, remained almost the same!

FLOC'H: That was not the concentration -- I have no idea of the concentration. It was % of radioactivity in the different compartments.

SRIVASTAVA: And that does not change in the conducting zone for 13 days!

SCHMITZ: If you looked only at % radioactivity, could it be that you have only a very tiny little bit of ^{32}P left in the source of the uptake region after 13 days and you compute that up to 100%.

FLOC'H: I doubt that. I measured that less than 1% of the radioactivity taken up was translocated per 24 hours.

SCHMITZ: Were those pulse-labelling experiments, or did you feed continuously for 13 days?

FLOC'H: They were pulse-labelled, of course.

SRIVASTAVA: Is it possible that a little bit of the inorganic phosphate is all the time being converted to organic form and being translocated.

FLOC'H: Well, what surprises me and what is evident is these large pools of ^{32}P in the uptake regions. It is the same in higher plants, for example for phosphate uptake by potato tuber cells. There is a slow but permanent transfer from the organic pool to the inorganic pool.

SCHMITZ: Did you analyze for polyphosphates?

FLOC'H: No. Not much is known about the localization of these inorganic pools -- localization in the cell or even in which cell.

DUNCAN: Did you see any difference in uptake rate or translocation rate or in products between the high-growing and subtidal algae?

FLOC'H: I have not compared the rates of translocation in algae in the higher intertidal and subtidal. But there is translocation in algae of each of these different levels -- in Fucus, Cystoseira and in Laminaria.

DUNCAN: Is there any difference in the compounds, such as polyphosphates? The reason I ask is that the amount of acid phosphatase is much higher in the high growing plants than in low growing plants, which should be related to the amount of phosphates inside or coming into the plant. Perhaps this might show in your work.

FLOC'H: Perhaps. I do not know.

TRANSLOCATION OF ORGANIC COMPOUNDS IN LAMINARIALES

K. Schmitz
Botanisches Institut der Universität zu Köln,
Gyrhofstr. 15, D-5000 Köln 41, F.R. Germany

Introduction

Two transport pathways may traditionally be distinguished in plants, the apoplastic and the symplastic (21). Symplastic transport seems advantageous for algae because it is more efficient and the translocate can better be conserved and kept under control. As far as we can see, only the symplastic pathway is used for the distribution of important nutrient materials in Laminariales.

A cooperative medium-distance (parenchyma-transport) and a long-distance transport of organic compounds are operational in the sporophytes of Laminariales. Photosynthetic CO_2-fixation is mainly confined to the meristoderm and the outer cortex of the thallus and the photoassimilate must move through the cortical cells before entering the long-distance transport system in the medulla (14,41). This parenchyma transport probably occurs via cytoplasmic intercellular connections, the plasmodesmata. The distribution and frequency of these connections was recently studied for two species of Laminaria (31). Although these connections are common and show up readily between meristoderm and the inner cortical cells, their occurrence between sieve elements and cortical cells is rare and so far has been seen only in the blades, not in the mature stipes of Laminariales. In stipes, the sieve element system seems to be "isolated" from other cells. If the sieve elements and cortical cells form a symplast, which seems to be the case in the blades, loading of the sieve elements is possible as long as there is a continuous concentration gradient from the meristoderm to the sieve tubes. This gradient has not yet been measured. The concentration difference between the sieve elements and the adjacent cortical cells is also not known. It

Synthetic and Degradative Processes in Marine Macrophytes

may well be that the concentration of solutes within the sieve elements, as indicated by an analysis of the sieve tube sap (25,37), is higher than in the cortical cells, which would necessitate an uphill transport mechanism. Thus, although there are structural indications for a symplastic transport of photosynthate, there are still no supporting physiological data.

Structural Aspects of the Long-Distance Transport System

Long-distance transport of photoassimilate and inorganic ions seems to be a common feature of Laminariales (8,32,42). All species investigated so far have been shown to have a long-distance transport system in the medulla. This system is composed of separate files of elongated sieve elements, which are inter-connected by other cross-connecting sieve elements. Compared to the inner cortical cells, the sieve elements have increased numbers and increased diameters of perforations in their cross walls, an anatomical feature which should considerably facilitate a symplastic transport along these cells.

Various genera of Laminariales show different degrees of specialization with respect to their sieve elements, from the most "primitive" sieve elements having small but numerous pores as in _Laminaria_, to those of _Macrocystis_ with relatively few but wide pores resembling the sieve tube members of _Cucurbita_. While the sieve pores of _Laminaria_ may still be regarded as plasmodesmata (35,40,46), the sieve pores of _Macrocystis_ closely resemble the sieve pores of higher plants (26,45).

In cross and radial longitudinal sections of stipes and blades the sieve elements show three successive developmental stages: - their origin from innermost cortical cells and differentiation as functional sieve elements, - the mature phase, when they function as transport cells, - their structural and functional degeneration.

A mature sieve element, therefore, is a cell structurally specialized to perform optimally as a transport cell. The sieve elements of the Laminariales are not associated with companion cells. Thus, there is no separation between a conducting compartment (the file of the sieve elements) and an energy-providing compartment (the companion cell), as in the phloem of angiosperms. Both functions are integrated in each sieve element of the Laminariales. This integration is reflected in their fine structure. Mature sieve elements of Laminaria and Alaria are characterized by numerous small vacuoles and the presence of cell organelles and membrane systems typical of most parenchyma cells (35,36,40). The sieve elements of Macrocystis, however, reveal a thin peripheral layer of cytoplasm with plastids, vesicles, mitochondria, occasional dictyosomes and endoplasmic reticulum (26,45), with the central part of the cell relatively free of cytoplasmic components (unpublished observations). According to Parker and Huber (26), however, the central lumen is filled with fine granular to fibrillar material or masses of electron-dense aggregated or dispersed slime depending on the method of fixation. In our own observations, P-protein has never been detected in the sieve elements of the Laminariales. Mature sieve elements of Macrocystis apparently lack a nucleus and a tonoplast as well, and their sieve pores appear as open intercellular connections.

Physiology of Translocation

The most specialized long-distance transport system in the Laminariales seems to be the one in Macrocystis. The following presentation, therefore, will mainly focus on this genus. The working hypothesis for the long-distance transport of organic compounds in Macrocystis is the osmotic pressure flow hypothesis originally proposed by Münch (21). The conditions for an osmotic pressure flow are:

- the existence of separate "source" and "sink" areas,
- the linkage of source and sink by a continuous file of conductive sieve elements with open sieve pores and unobstructed lumina,
- an enclosure of the transport compartment by a semipermeable membrane,
- a gradient of osmolarity between source and sink which must create a parallel pressure gradient high enough to drive a solution flow,
- removal of osmotically active translocate from the transport compartment at the sink,

- efflux of water from the transport system, mainly at the sink.

The consequences of such transport are:

- a strictly directed transport from source to sink,
- a mass stream of dissolved matter and the solvent water,
- movement of different substances at the same time in the same direction,
- movement of different substances at the same speed unless electrostatic or metabolic factors are involved.

The existence of source and sink areas in the Laminariales was convincingly demonstrated by experiments with intact and amputated plants and by measurements of growth and rates of carbon fixation (2,3,7,17,18,22,30). These experiments showed that growing regions were carbon deficient (sinks) and dependent, for normal growth, on carbon supply from surplus producing areas (sources) such as mature blades.

That organic compounds are translocated in the Laminariales was demonstrated by use of radioisotopes, mainly ^{14}C. But dyes (24,25) and other radioisotopes, organically bound or as inorganic ions, have also been shown to be mobile in these plants (Table I). The transport of these isotopes and dyes is always directed toward the sink areas. More recent experiments using ^{14}C to label the photosynthate in the blades have confirmed that photoassimilate moves from the ^{14}C-exposed blades along the stipe of Macrocystis to the growing regions (15,16,24,32,37,38). Movement of ^{14}C along a stipe occurs in intact plants as well as in exuding cut fronds which release the radiocarbon with the sieve tube sap (25,37,38).

Radioactive translocate has been localized by histoautoradiography within the sieve elements along the transport pathway (38,41). Moreover, cut stipes of Macrocystis exude a clear fluid at a rate of 0.5 - 2 ml h^{-1} or more, and in experiments using ^{14}C the exudate contains ^{14}C-labelled organic compounds (6,25,34,37,38). The exudate is released by the mature sieve tubes at the periphery of the medulla, cells which were identified by histoautoradiography to be the translocating sieve elements (38). From these facts, we conclude that the exudate is the sieve tube sap and represents the translocate, and furthermore, that the source and sink regions are connected by a continuous conductive long-distance transport

Table I: Species of brown algae that have been shown to translocate various radioisotopes.

Plant species	Radioisotope	Reference
Chordaceae		
Chorda filum	^{14}C	(42)
Laminariaceae		
Laminaria digitata	^{14}C, ^{32}P, ^{86}Rb, ^{35}S	(9,10,11,14)
L. hyperborea	^{14}C, ^{32}P, ^{86}Rb	(12,19,33,41)
L. japonica	^{14}C	(42)
L. saccharina	^{14}C	(19,20,33)
L. setchellii	^{14}C	(32)
L. sinclairii	^{14}C	(32)
Pleurophycus gardneri	^{14}C	(32)
Cymathere triplicata	^{14}C	(32)
Costaria costata	^{14}C	(32)
Agarum fimbriatum	^{14}C	(32)
Hedophyllum sessile	^{14}C	(32)
Lessoniaceae		
Postelsia palmaeformis	^{14}C	(32)
Nereocystis luetkeana	^{14}C	(23,32)
Macrocystis integrifolia	^{14}C, ^{32}P, ^{3}H	(16,32,37,38,39)
M. pyrifera	^{14}C	(15,24,25)
Lessoniopsis littoralis	^{14}C	(32)
Alariaceae		
Alaria esculenta	^{14}C, ^{32}P, ^{86}Rb	(2,10)
A. marginata	^{14}C	(32,36)
Egregia laevigata	^{14}C	(32)
Eisenia arborea	^{14}C	(32)
Fucaceae		
Fucus vesiculosus	^{32}P, ^{86}Rb	(10)
Ascophyllum nodosum	^{32}P, ^{86}Rb	
	^{45}Ca, ^{36}Cl, ^{99}Mo	(29)
Sargassaceae		
Sargassum pallidum	^{14}C	(42)

system. These facts also suggest that the structure of the sieve tubes of Macrocystis is conducive to a solution flow, which would probably not occur if the sieve pores were occluded or the lumina of the sieve elements filled with dense cytoplasmic components.

A comparative analysis of the photosynthate and translocate revealed that nearly the same compounds which are radioactively labelled during photosynthesis are subject to long-distance transport. The main labelled carbohydrate is mannitol. In addition, hexose- monophosphates, -diphosphates, PGA, and several amino acids are also heavily labelled. The percentage of radioactivity confined to the amino acids in the translocate may be as high as 40%. The chemical analysis of the sieve tube sap shows that mannitol concentration may be as high as 10% (w/v) or 67% (w/w) on a dry weight basis. Amino acids comprise 15% (w/w) and potassium roughly 8% (w/w) on a dry weight basis. Sodium, on the other hand, may be as low as 0.75 to 0.8 mg ml^{-1} of the sieve tube sap. These data indicate that the most heavily labelled substances, mannitol and amino acids, are also quantitatively the principal substances that are translocated. In addition, the concentration of potassium in the sieve tube sap is much higher than in the sea water, whereas the reverse is true for sodium. This ionic selectivity clearly suggests that the plasmalemma of the transport compartment (sieve tubes) is a selectively permeable barrier.

The release of appreciable amounts of exudate for several hours from cut ends of Macrocystis stipes can only be explained if it is assumed that the sieve elements are under high hydrostatic pressure. A continuous loading of osmotically active substances into the sieve elements in the source region would cause water to be attracted to the sieve elements and result in the build-up of pressure high enough to drive the solution along the sieve tubes to the cut end of stipe (sink). This phenomenon can be demonstrated by using frond cuttings, 60 to 100 cm long with 3 to 6 attached blades, which are still able to release exudate with a uniform rate for several hours. If radiocarbon is continuously supplied to a mature blade under conditions of photosynthesis, ^{14}C will normally be released with the exudate. The time interval between the feeding of ^{14}C to the blade and its appearance in the sieve tube sap at the cut end of the stipe is a rough measure of the translocation velocity. Similar experiments may be performed using the radioisotopes ^{32}P and ^{3}H, fed singly or in combination with ^{14}C to the same blade. The data from such an experiment are presented in figure 1 and Table II and show that ^{14}C and ^{32}P moved upward and downward along the stipe. Most radioactivity was, however, translocated toward the basal cut end of the stipe. Radioactivity first appeared in the sieve tube sap 90 to 100 minutes after the experiment was started and steadily increased up to 240 minutes when the experiment was stopped. Since the fed blade and the site of sieve tube sap collection were 680 mm apart, transport velocity must have been at least 41 to 45 cm h^{-1}. This experiment also shows that ^{14}C and ^{32}P when fed simultaneously are translocated together in the same direction. It also suggests that the radioactively labelled compounds in the sieve tube sap, identified by thin-layer chromatography, move in a solution stream (see also references 37,38).

If translocation of photoassimilate in Macrocystis is a solution stream, the main translocate should be the solvent water. Provided there is no channelized recirculation of water, water should enter the sieve element system in the blade (source) following the import of osmotic substances and it should leave the sieve tubes mainly at the sink. Several research groups have tried to demonstrate the transport of water in the phloem of vascular plants using tritiated water (THO), but the results have been mostly contradictory (1,4,5,13,27,28,43,44). We have tried to trace the movement of THO in M. integrifolia by applying THO to a mature blade under conditions of photosynthesis. Such an application of THO has the advantage that it yields ^{3}H-labelled water and ^{3}H-labelled photosynthate, both of which can be analyzed separately. The experiments were done with intact fronds as well as frond segments which released exudate. The movement of THO and ^{3}H-labelled organic compounds occurred in both systems. An experiment with an intact frond is presented in figure 2. A blade close to the growing apex was exposed to a THO-seawater medium for 3 h. The distribution of THO and ^{3}H-labelled organic compounds along the stipe after 3 h shows that radioactivity moved nearly exclusively toward the apical growing region where a slight accumulation of ^{3}H-label occurred (39). Experiments with frond segments also revealed a movement of ^{3}H

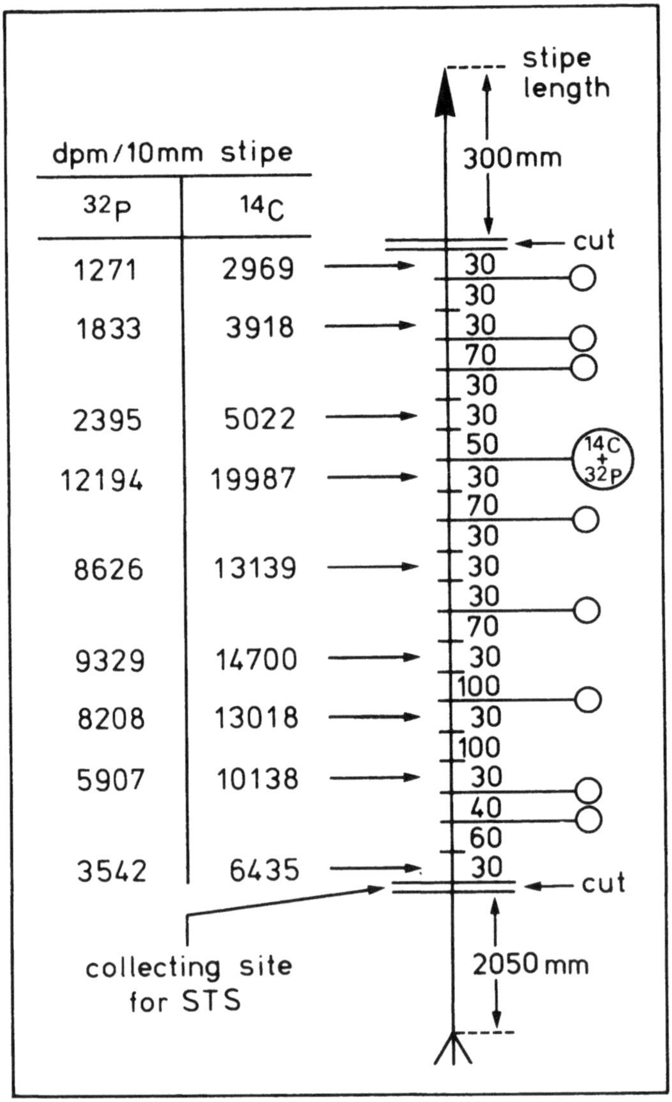

Fig. 1: Translocation of ^{14}C and ^{32}P along a stipe of a frond cutting of Macrocystis integrifolia. The indicated blade of the intact frond was exposed for 4 h to a sea water medium containing ^{14}C- sodium bicarbonate and ^{32}P- orthophosphate under conditions of photosynthesis. Fifteen minutes after the experiment was started, the apical and basal parts of the frond were cut off. Sieve tube sap was collected from the basal end in 10 min batches for the duration of the experiment and analysed for ^{14}C and ^{32}P (see Table II). At the end of the experiment, 30 mm pieces of stipe were sampled both apically and basally from the fed blade, extracted and counted for radioactivity. The dpm/10mm stipe for both ^{14}C and ^{32}P in the stipe segments are listed.

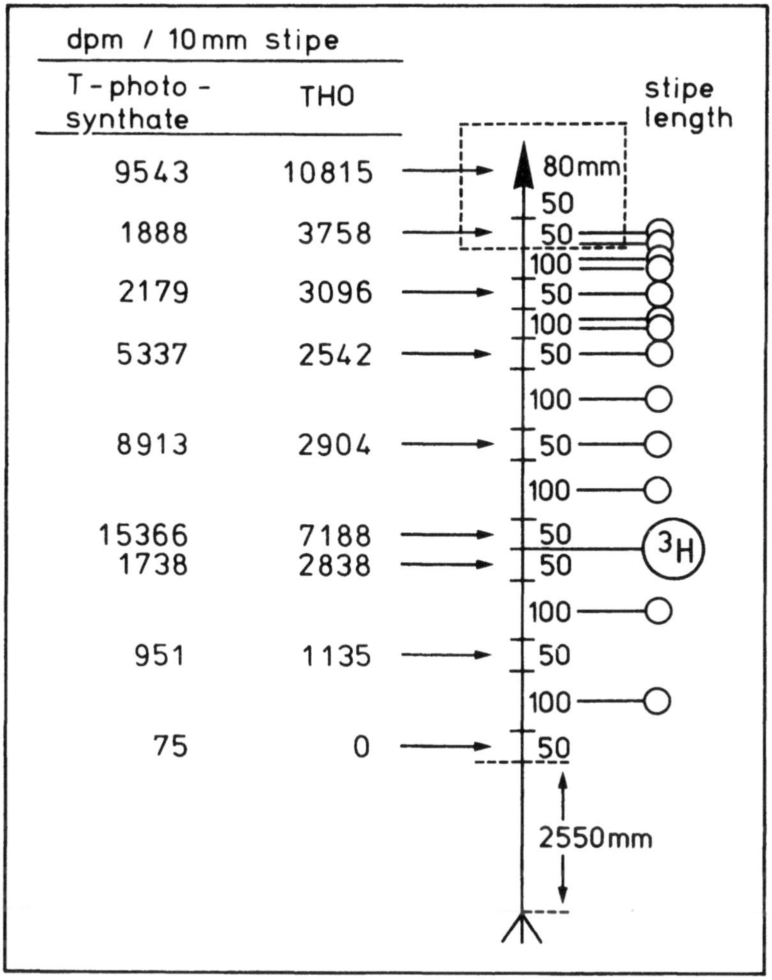

Fig. 2: Translocation of ^{3}H in an intact frond of Macrocystis integrifolia. A mature blade was exposed to a THO-seawater medium for 3 h in the light. The growing apex was enclosed in a translucent plastic bag with 200 ml of sea water enriched with carbonate. At the end of the experiment 50 mm pieces of stipe were rapidly cut at 100 mm intervals and immediately deep frozen. These samples were lyophilized, the condensate from each sample was collected separately and the radioactivity, confined to THO only, was measured. From the dried tissue samples, ^{3}H-labelled organic compounds were extracted and their radioactivity measured. The data are presented in the figure. A water sample was withdrawn from the plastic bag enclosing the apex and lyophilized. The condensate revealed appreciable radioactivity of THO (for details see text).

Table II: Release of ^{14}C and ^{32}P with sieve tube sap (STS) collected from a frond segment shown in figure 1. STS was collected in 10-min fractions.

Experiment time (min)	dpm/ 25 µl STS ^{14}C	^{32}P
90 - 100	27	0
100 - 110	48	10
110 - 120	148	72
170 - 130	210	131
130 - 140	221	204
140 - 150	254	363
150 - 160	479	467
160 - 170	528	817
170 - 180	1086	1119
180 - 190	1254	2135
190 - 200	1558	1973
200 - 210	1750	1954
210 - 220	1792	2129
220 - 230	2034	2739
230 - 240	2059	3732

along the stipe towards the basal cut end. Both THO and ^3H-labelled compounds were detectable and showed a clear gradient in extracts of stipe sections cut at regular intervals along the transport pathway at the end of an experiment. Tritiated label was also released with the exudate collected at the cut end of the stipe; the ^3H-radioactivity in the exudate steadily increased with time when THO was fed continuously to a mature blade about 50 cm away. Nearly all the radioactivity in the sieve tube sap was confined to organic compounds; the radioactivity of THO was always low and did not change significantly with time, which suggests that THO rapidly exchanges with water in the surrounding tissue and perhaps with the sea water outside the plant (39, but cf. 4,27,28,44). If THO is indeed transported along with ^3H-labelled organic compounds, there should be a release of THO at the sink area (growing apex) as well. The apex of a M. integrifolia frond, therefore, was enclosed in a plastic bag with 100 to 200 ml of sea water enriched with carbonate and optimized in oxygen (see Fig. 2). At the end of the experiment, after 3 h, the sea water in the bag was analysed for THO and revealed more than 2,500 dpm of $^3H\ ml^{-1}$

of distillate which clearly indicates that THO is released from the transport compartment at the sink.

Concluding Remarks

The Laminariales have evolved an effective long-distance transport system for inorganic and organic compounds. The anatomy and fine structure of sieve elements in different species shows that they have reached different levels of specialization, some clearly more advanced than the others. In an evolutionary line, *Laminaria* would perhaps represent the most primitive and *Macrocystis* the most specialized sieve tube system. There are also several structural similarities between the sieve elements of the Laminariales, especially *Macrocystis*, and those of vascular plants, but a number of unique features such as the presence of vesicles and organelles, including a nucleus, underscore the suggestion that the sieve elements of the Laminariales have evolved parallel to those of higher plants. The experimental data on the physiology of translocation in *Macrocystis*, which apparently has the most specialized transport system, support a pressure-flow mechanism for the movement of water (solvent) and dissolved organic and inorganic materials. This solution moves in a strictly directed manner in the sieve elements which connect source and sink areas. The transport system is loaded with osmotic substances (assimilates) at the source region or mature blades, which causes water to enter the sieve elements and later both, the water and the dissolved osmotic materials, are removed from the translocation pathway at the sink (growing regions). We assume that the driving force for the long-distance transport through the sieve elements might be an osmotic gradient along the transport pathway. The existence of such an osmotic gradient still remains to be shown.

References

1. Biddulph, O., Cory, R.: Plant Physiol. 32, 608-619 (1957).
2. Buggeln, R.G.: J. Phycol. 10, 283-288 (1974).

3. Buggeln, R.G.: J. Phycol. 13, 212-218 (1977).

4. Cataldo, D.A., Christy, A.L., Coulson, C.L.: Plant Physiol. 49, 690-695 (1972).

5. Choi, I.C., Aronoff, S.: Plant Physiol. 41, 1119-1129 (1965).

6. Crafts, A.S.: Am. J. Bot. 26, 172-176 (1939).

7. Duncan, M.J.: Helgoländer wiss. Meeresunters. 24, 510-525 (1973).

8. Floc'h, J.Y., Penot, M.: C.R. Acad. Sci., Paris 271, 288-291 (1970).

9. Floc'h, J.Y., Penot, M.: C.R. Acad. Sci., Paris 273, 1100-1103 (1971).

10. Floc'h, J.Y., Penot, M.: Physiol. Vég. 10, 677-686 (1972).

11. Floc'h, J.Y., Penot, M.: C.R. Acad. Sci., Paris 282, 989-992 (1976).

12. Floc'h, J.Y., Penot, M.: Z. Pflanzenphysiol. 96, 377-385 (1980).

13. Gage, R.S., Aronoff, S.: Plant Physiol. 35, 53-64 (1960).

14. Hellebust, J.A., Haug, A.: Can. J. Bot. 50, 169-176 (1972).

15. Lobban, C.S.: Plant Physiol. 61, 585-589 (1978).

16. Lobban, C.S.: J. Phycol. 14, 178-182 (1978).

17. Lüning, K.: Mar. Biol. 2, 218-223 (1969).

18. Lüning, K.: Helgoländer wiss. Meeresunters. 20, 79-88 (1970).

19. Lüning, K., Schmitz, K., Willenbrink, J.: Proc. Int. Seaweed Symp. 7, 420-425 (1972).

20. Lüning, K., Schmitz, K., Willenbrink, J.: Mar. Biol. 23, 275-281 (1973).

21. Münch, E.: Die Stoffbewegungen in der Pflanze. Verlag G. Fisher, Jena 1930

22. Nicholson, N.L.: J. Phycol. 6, 177-182 (1970).

23. Nicholson, N.L., Briggs, W.R.: Am. J. Bot. 59, 97-106 (1972).

24. Parker, B.C.: J. Phycol. 1, 41-46 (1965).

25. Parker, B.C.: J. Phycol. 2, 38-41 (1966).

26. Parker, B.C., Huber, J.: J. Phycol. 1, 172-179 (1965).

27. Peel, A.J.: Physiol. Plant. 23, 667-672 (1970).

28. Peel, A.J., Field, R.J., Coulson, C.L., Gardner, D.J.C.: Physiol. Plant. 22, 768-775 (1969).

29. Penot, M., Penot, M.: Phycologia 16, 339-347 (1977).

30. Sargent, M.C., Lantrip, L.W.: Am. J. Bot. 39. 99-107 (1952).

31. Schmitz, K., Kühn, R.: Planta (in press).

32. Schmitz, K., Lobban, C.S.: Mar. Biol. 36, 207-216 (1976).

33. Schmitz, K., Lüning, K., Willenbrink, J.: Z. Pflanzenphysiol. 67, 418-429 (1972).

34. Schmitz, K., Srivastava, L.M.: Planta 116, 85-89 (1974).

35. Schmitz, K., Srivastava, L.M.: Cytobiology 10, 66-87 (1974).

36. Schmitz, K., Srivastava, L.M.: Can. J. Bot. 53, 861-876 (1975).

37. Schmitz, K., Srivastava, L.M.: Plant Physiol. 63, 995-1002 (1979).

38. Schmitz, K., Srivastava, L.M.: Plant Physiol. 63, 1003-1009 (1979).

39. Schmitz, K., Srivastava, L.M.: Plant Physiol. 66, 66-69 (1980).

40. Sideman, E.J., Schreier, D.C.: Am. J. Bot. 64, 649-657 (1977).

41. Steinbiss, H.H., Schmitz, K.: Planta 112, 253-263 (1973).

42. Titlyanov, E.A., Peshekhodko, V.M.: Trudy Biologo-počvennogo instituta Novaja serija 20, 137-140 (1973).

43. Trip, P., Gorham, P.R.: Plant Physiol. 43, 1845-1849 (1968).

44. Tyree, M.T., Zimmerman, M.H.: In Encyclopedia of Plant Physiology, New Ser. 1, (M.H. Zimmermann, J.A. Milburn, eds.), Springer, Berlin, pp. 478-479 1975

45. Ziegler, H.: Protoplasma 57, 786-799 (1963).

46. Ziegler, H., Ruck, I.: Planta 73, 62-73 (1967).

DISCUSSION

DRUEHL: Have you looked at day vs. night translocation?

SCHMITZ: No.

DRUEHL: Is there any possibility of an experiment like that? It might give you an idea of the osmotic potential as a mechanism for translocation.

SCHMITZ: My ideas about finding out whether there is osmotic potential go in another direction. I think one has to prepare a plant, just leave one blade, cut the others off, and then analyze stipe sections for a gradient in radioactivity. I have looked at the sieve tube sap; if you leave the other blades on you will not be able to pick up a clear gradient.

DRUEHL: In what stage of development is the nucleus lost and is the nucleus absent for most of the species for which there is translocation?

SCHMITZ: I am pretty sure that *Macrocystis* sieve elements do not contain a nucleus at maturity. We have tried Feulgen staining and serial sections and we have not found any nuclei in mature elements. When the nucleus is lost, the central part of the sieve element is nearly empty and the sieve pores have reached their largest size, also they are open. That is in *Macrocystis*. In *Laminaria* we are looking at totally different sieve elements -- they are not so well developed, they contain lots of vacuoles still, they contain chloroplasts and mitochondria and they contain nuclei. So you have a developmental line in an evolutionary sense from let us say *Hedophyllum*, which looks even more primitive than *Laminaria*, to *Macrocystis*. You will find the same situation in a *Macrocystis* plant also, if you look at the sieve elements, from the inner cortex toward the medulla; you have the whole series, which you have in the phylogenetic sense, in one and the same plant. *Macrocystis* just goes further than *Laminaria* in modification of the sieve element protoplast.

SRIVASTAVA: On the question of the nucleus, Mao Lien, a graduate student of mine, has some stages in the dissolution of the nucleus in *Macrocystis*. So most likely it does go.

FLOC'H: I have a comment on uptake and our controversy about the form in which phosphorus is translocated. Actually, the controversy is only apparent. Our results are quite similar, but we have two possible interpretations. We have no definite conclusions or proof yet as to whether phosphate is translocated as inorganic phosphate or as hexose-monophosphate. I think it would be important to find out which substance enters first into the translocating cells, but that would be very difficult to solve, because of the very fast turnover of these phosphorylating compounds. The turnover of ADP is about one minute, and that of the hexose-monophosphates is about 5-6 minutes. So I think the problem is very hard.

SCHMITZ: I think we were very careful in pointing out that our data suggest that phosphorus is being translocated in the organically-bound form, but we do not want to exclude the possibility that it may also be translocated as phosphate. You know that we did *in vitro* incubation of sieve tube sap with radioactive phosphate and found a rapid incorporation of phosphate into many substances. So the sieve tube sap does contain the enzymes for these syntheses. I also made it clear that the sieve element in the Laminariales are different from those of vascular plants because we have here everything in one cell. Here, it appears, we do not have the separation of the translocating part from the metabolically active part, the companion cells which are nucleate and packed with mitochondria and ribosomes, as in vascular plants.

FLOC'H: That was what I meant. I have another question. It is about the direction of translocation. Have you found a competition between the holdfast and scimitar in <u>Macrocystis</u>?

SCHMITZ: Yes, there is competition along the stipe. If the blade is closer to the tip, most of the translocate will move to the tip, and if the blade is closer to the holdfast, most of the translocate will move to the holdfast. If there are young fronds at the holdfast, they will attract even more translocate. If the blade is in an intermediate position, you will get transport up and down. We do not exactly know how this works, but it does.

BIDWELL: I was very interested in your data where the plant was translocating towards the base, after it had been cut off. If the system was working exclusively with a Münch type pressure flow system, of course as soon as you cut it off, the translocation ought to stop, but this does not happen. It is the same in higher plants; you can cut the plant off from its supposed sink and the translocation will still continue. The inference is, that there may be unloading mechanisms involved that are affected other than just the avidity of the sink for the solutes that are present.

SCHMITZ: I think that the fact of cutting creates an enormous sink and the solution just flows out, and that is the unloading process. But if the whole system is closed, we have to postulate that the translocate as well as the water have to move out, otherwise, the osmotic capacity will build up to the point that the whole thing comes to a stop. Also, the water has to come out.

BIDWELL: Well, if you had an unloading at the sink, then the water would follow the unloading by osmosis, would it not?

SCHMITZ: Yes. It will follow passively.

SRIVASTAVA: Going back to the question of directionality, there may be something to the concept that Klaus does not mention, that is of "imprinting." The conduits get imprinted toward flow in a certain direction for sometime, and then something else has to happen, which is probably metabolically controlled, to change the directionality. Even in these plants, the apical blades, which are for a long time transporting upwards only, towards the close of the fall when the scimitar has matured, the photoassimilate travels primarily downwards. So there is a change in directionality there.

BIDWELL: This business of the upper fronds translocating upwards and the lower fronds translocating downwards that is all fact. The question has been examined, and if the system is an absolutely open tube pressure flow system the proportion going to any sink should be the same for every leaf from the system, irrespective of the distance, because the proportional removal from different sinks is the same. You find exactly the same situation in virtually all higher plants where the flagleaf feeds the seed and in wheat where the lower leaves feed the root. This is presumably controlled by some additional mechanisms.

SCHMITZ: I suppose if we could perform an experiment, collecting from the basal and the apical cuts simultaneously and labelling the blade exactly in the middle, we would obtain the same radioactivity. I do not know? We expect it! As to the directionality of ^{32}P transport, it is a transport against a concentration gradient. I think the directionality in these plants is given by the overall concentration gradient in the sieve tube system and that will probably direct the flow.

Part III. Marine macrophytes in coastal ecosystems

THE REGULATION OF MACROALGAL ASSOCIATIONS IN KELP FORESTS

M.S. Foster

Moss Landing Marine Laboratories
P.O. Box 223, Moss Landing, CA 95039, U.S.A.

Introduction

What processes control or regulate macroalgal species composition and abundance in kelp forests? The answer to this question has been approached for other communities (7), particularly the rocky intertidal zone, where the organisms are accessible, well described and relatively easy to manipulate (6). In contrast, kelp forests are relatively inaccessible, difficult to manipulate and most studies on the Pacific coast of North America have been confined to southern California (5,10,16,30,32). Moreover, much of this work involved qualitative descriptions made during the calmer summer months (for an exception, see Rosenthal et al. [32]). These and other studies (3,12,25,26,31) do, however, document seasonal fluctuations in algal abundance within a particular forest, as well as dramatic spatial changes in abundance and species composition under the same canopy type.

A number of factors have been suggested as important in regulating these algal associations on a local scale (eliminating geographic variation). Sand may prevent settlement or destroy established plants by burial and abrasion (3,17,32). Detachment mortality may be dependent on substratum hardness. Light affects a variety of plant processes and can vary with turbidity, depth, season and overstory canopy development (e.g., algal-algal competition for light [5,10,17,27,31,32]). Nutrients undoubtedly affect growth as well (21). Water motion can affect settlement (28), remove adults (3,5,32,33) and alter nutrient availability (21,28). Grazing can modify or completely remove vegetation (3,5,10,14,17,23,31), and competition with other algae (11,16,17) and sessile invertebrates (15,17,29) for space can affect species composition

Synthetic and Degradative Processes in Marine Macrophytes

and distribution. Many of these factors interact and others could be added. Their relative importance varies from site to site, but at any one site probably several factors play a role.

What are the dominant regulatory processes for kelp forests and how do they vary from one forest to another? This paper attempts to answer these questions for three kelp forests in central California. Long-term seasonal changes in algal species composition and abundance are correlated with site-specific physical and biological characteristics of apparent importance to the organization of algal associations. These correlations are explained by hypotheses that are partially tested by manipulative field experiments and site contrasts.

Study Areas

Six kelp forests were investigated over the past three years, but only three are discussed in detail here. A complete discussion of all sites and methods is available in Foster et al. (18). These three forests, designated as Stillwater Cove, Sandhill Bluff and Greyhound Rock, are located between Carmel Bay and Año Nuevo Island in central California (Fig. 1). The coastal climate in this region is highly seasonal, with most large swells associated with winter storms. Upwelling is common in the area and nearshore nutrient levels are generally high (4). Nearshore temperatures (4) and on-site temperature measurements using diver-held thermometers reveal no significant differences between sites, and salinities are also similar (4).

The physical and biological characteristics of the sites are summarized in Table I. Stillwater Cove is relatively protected from the predominantly northwest winter swell; it is inside Carmel Bay and opens to the south. The substratum in both the kelp forest and the intertidal zone is hard rock of moderate relief. Diver observations and secchi disc readings indicate long periods of relatively clear water. Macrocystis pyrifera (giant kelp) is the most common surface canopy kelp and grows from depths below 30 m.

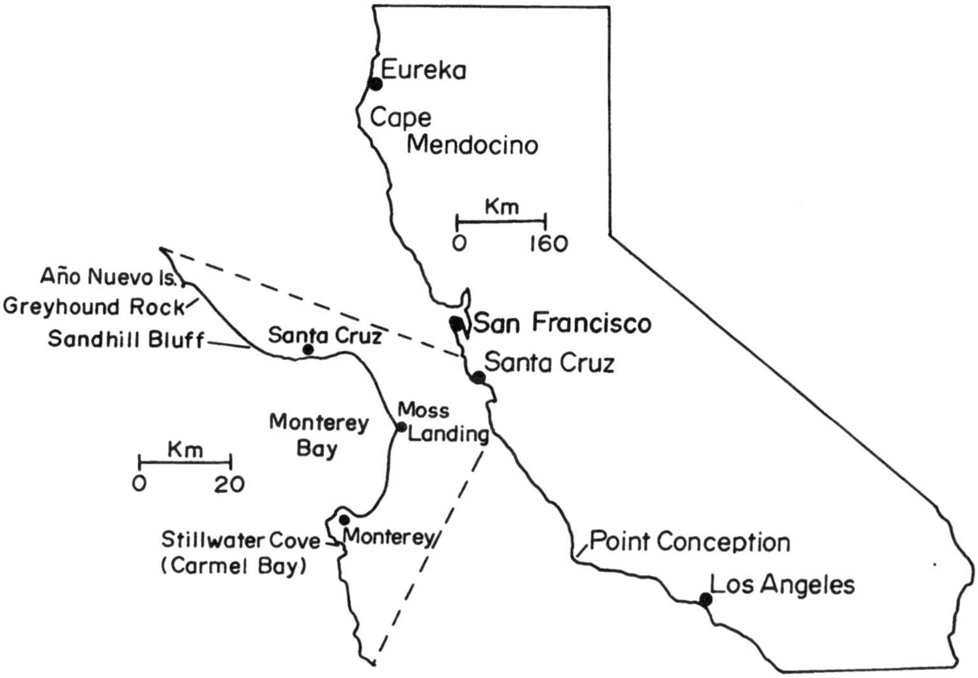

Fig. 1: Survey site locations.

The shore north of Santa Cruz is generally sandy beach backed by high cliffs. The substratum is sand and/or Santa Cruz mudstone with some shale to depths of 20 m. This rock varies in hardness, but is generally soft and easily eroded. The subtidal substratum at Sandhill Bluff is generally flat, while Greyhound Rock is composed of moderate-relief parallel ridges. The coast in this region faces southwest and, except for the area immediately southeast of Año Nuevo Island, is fully exposed to northwest swells. These swells frequently exceeded a height of 4 m between October and April, 1977-78, and breaking surf was observed over site buoys anchored at 10 m depth. Swell during the winter of 1978-79 was more moderate (generally less than 3 m). The Sandhill Bluff site is partially protected by a small point and faces more to the southwest so water motion is slightly reduced relative to Greyhound Rock. High water motion disturbs the nearshore sediments and, combined with run-off from numerous small creeks, produces generally turbid water from late fall through early spring at these northern sites. Macrocystis pyrifera forms the surface

Table I: Physical and biological characteristics of the survey sites.

Character	Stillwater Cove	Sandhill Bluff	Greyhound Rock
Depth	9-14 m	6-9 m	6-12 m
Substratum[a]	sandstone	mudstone	mudstone
Topographic Relief[b]	moderate	low	moderate
Water Clarity[c]	clear	turbid	turbid
Swell Exposure	low	moderate	high
Grazers[d]	small snails and crabs	urchins approx. $1/m^2$	urchins approx. $3/m^2$
Canopy Kelp	*Macrocystis pyrifera* to 30 m	*Macrocystis pyrifera* to 12 m	*Nereocystis luetkeana* to 12 m
Understory[e]	*Pterygophora californica* and coralline algae	fleshy red algae and *Desmarestia ligulata*	fleshy red algae and *Desmarestia ligulata*
# of species[f]	14 (13-15)	31 (29-33)	30 (24-34)

[a] Stillwater Cove also has some conglomerate and lava.
[b] Qualitative estimates from surveys; low 1-2 m; moderate 2-5 m.
[c] Average qualitative estimates made during all surveys based on diver estimates and secchi disc measurements (1-5 m - turbid; >8 m - clear).
[d] Composition and abundance determined by counts of plants and in quadrats.
[e] Only the most common species and types listed.
[f] Some groups, such as encrusting corallines, not determined to species. Data are means for summer surveys (range).

canopy at Sandhill Bluff, but does not grow below 12 m, even though suitable substratum is available. Greyhound Rock is a *Nereocystis luetkeana* forest. This canopy kelp is also rarely found below 12 m.

Methods

Permanent, long-term survey areas were established at Sandhill Bluff and Greyhound Rock in summer (July) 1977 and at Stillwater Cove in early spring (March) 1978. These areas have been surveyed in spring (March-May) and summer (July-September) since that time. The survey areas at Greyhound Rock and Sandhill Bluff are circular (40 m radius), each encompassing 5026 m^2. The centers of these areas were originally placed at the seaward edge of the kelp surface canopies so that one-half of the survey site would be inside the canopy and one-half outside. After the initial surveys, it was apparent that the canopy edge varied due to continued growth or removal by storms. However, areas toward shore do have generally more adult kelp plants. Since giant kelp grows to depths below 30 m in Stillwater Cove, surveys there were done entirely within the forest. The data were collected in either one 40-m radius area of moderate depth (10-13 m) or two 20-m radius areas, one in shallow water (9-12 m) and one in deep water (12-14 m). The vegetation is similar over this entire depth range, so the different survey areas are not distinguished below.

Seasonal changes in surface canopy areas and density at the two Macrocystis pyrifera sites were determined from infrared aerial photographs taken by the California Department of Fish and Game. The total canopy cover in fixed locations, which included the survey areas, was determined from the projected transparencies. Subtidal plant cover was estimated by divers using a modification of the point contact method (16) employed by terrestrial plant ecologists (20). The sampling device consisted of a 1.4 m long metal bar with a 1.8 m long string attached to the bar like a loose string on a bow. Five knots were tied at random on the string. The bars were positioned at random within each study area. By moving the string from one side of the bar to the other, ten points were sampled per bar location. A point is determined by grasping a knot, pulling the string taut and pressing the knot to the substratum. The diver then notes the presence of all organisms intersecting an imaginary line extending through the knot perpendicular to the substratum to a height of 2 m. More than one alga can occur over and/or under one point, so algal cover can exceed 100%. Total cover of a particular species was calculated by dividing the sum of all contacts with the species (total number of points at which the species was found) by the total number of points surveyed. Algal species were identified using the monograph by Abbott and Hollenberg (1).

The abundance and distribution of large invertebrate grazers were determined at all three sites by counts in randomly-placed 1 or 5 m^2 quadrats and on Macrocystis pyrifera plants. Sea urchin (Strongylocentrotus franciscanus) removal experiments were performed at two of the sites. The methods and results of these experiments are available elsewhere (9,18), but some results will be mentioned in the following.

Results

The most abundant species throughout the year in the algal association at Stillwater Cove are Macrocystis pyrifera, a subcanopy of Pterygophora californica and a bottom cover of Calliarthron tuberculosum and various encrusting corallines. Cystoseira osmundacea is also common, especially in summer, when reproductive P. californica plants are large (1-1.5 m tall) and, if cortical rings are annual (19), over ten years old. The C. tuberculosum plants are also generally large (5-15 cm tall) and, given their slow growth rates (22), probably quite old.

Sandhill Bluff also has a surface canopy of Macrocystis pyrifera, but the subsurface canopy of Pterygophora californica is very sparse and plants are rarely over 0.5 m tall. Summer bottom cover is made up of fleshy red algae, especially Polyneura latissima, Plocamium cartilagineum and Phycodrys setchellii. Desmarestia ligulata var. ligulata is occasionally very abundant and encrusting corallines are common.

Subcanopy kelps and articulated corallines are also sparse in the Nereocystis luetkeana forest at Greyhound Rock. Bottom cover at this site is similar to Sandhill Bluff, with Polyneura latissima most abundant in summer, followed by Botryoglossum farlowianum var. farlowianum.

Desmarestia ligulata var. ligulata is abundant every summer, and encrusting corallines are common at all times of the year. The number of algal species within the survey areas is similar at the sites north of Santa Cruz. The number is consistently lower at Stillwater Cove (Table I).

While the patterns described above are generally true, there can be large seasonal and year-to-year variations in species composition and especially abundance. The summer of 1977 followed two years of very reduced winter storms and the Macrocystis pyrifera canopies at that time probably represent near-maximum development (Fig. 2). Canopy cover declined to near zero at both M. pyrifera sites by spring 1978, after a rather severe winter in 1977-78. The canopy at Stillwater Cove recovered more rapidly than that at Sandhill Bluff (Fig. 2), and although good quality aerial photographs were not available for spring 1979, direct observations indicated that minimum cover followed the winter storm period and maximum cover occured prior to the first storms in early fall. Our observations of other M. pyrifera canopies indicate this seasonal pattern is typical for central California. The canopy at Sandhill Bluff recovered more slowly from the winter storms of 1977-78, and was still absent over the survey area in summer 1978. Although the canopy at Sandhill Bluff exhibited a general spring minimum, fall maximum trend, the year-to-year differences were more pronounced at this site than at Stillwater Cove.

The Nereocystis luetkeana canopy at Greyhound Rock is thin and patchy, even at maximum development, and difficult to assess with aerial photography. Qualitative observations from sea and shore indicate it also reaches maximum cover in fall, although usually a month or two later than the Macrocystis pyrifera maximum. Maximum cover so far observed occurred in fall 1978. Minimum cover in this annual plant also occurs in spring.

Upright algal cover follows seasonal trends similar to the surface canopy kelps (Fig. 2). At Stillwater Cove, the cycle has been consistent, with a minimum of around 100% cover and a maximum of about 150%. In contrast, year-to-year understory cover at Sandhill Bluff has been highly

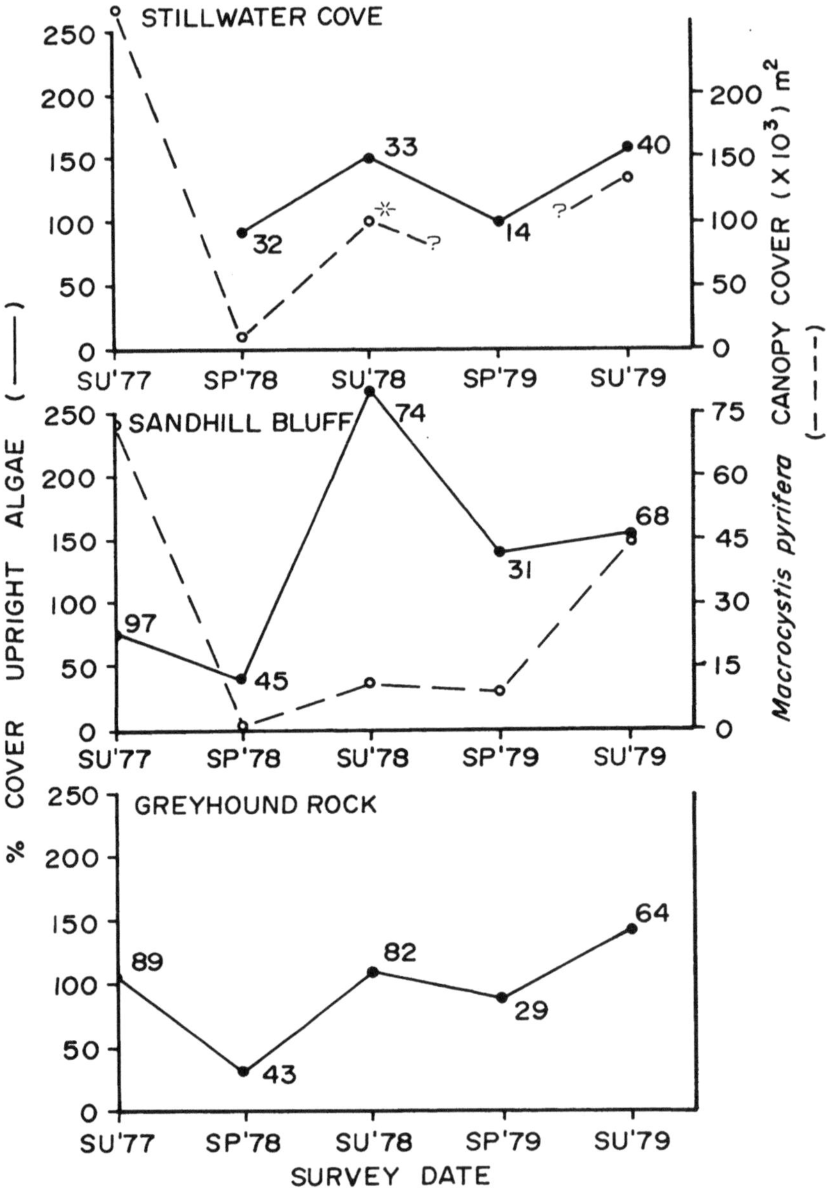

Fig. 2: Macrocystis pyrifera and upright (nonencrusting) algal cover at Stillwater Cove, Sandhill Bluff and Greyhound Rock. Mean percentage cover of upright algae and canopy cover (m^2) of M. pyrifera are shown. For upright algae, numbers indicate the number of groups of ten points in each sample; * = cover only approximate due to poor quality photo; ? = no photo available; su = summer; sp = spring.

variable with a maximum of 270% in the summer of 1978, about three times greater than in summer 1977. Similar year-to-year differences have been observed in the two other Macrocystis pyrifera forests surveyed north of Santa Cruz. Greyhound Rock is most similar to Sandhill Bluff in species composition, but understory algal abundance has not undergone the large year-to-year variation found at the latter site (Fig. 2). The variation in abundance at Greyhound Rock is more similar to Stillwater Cove, and a similar trend has been observed in the one other Nereocystis luetkeana forest surveyed about 2 km south of Greyhound Rock.

Figure 3 shows the variation in abundance of understory algae which are most common in at least one of the survey areas. Calliarthron spp. (primarily C. tuberculosum) and Bossiella spp. (B. californica and B. orbigniana) have been combined to represent the most abundant articulated corallines. The figure illustrates the differences in species composition between sites discussed above, as well as the relative contribution of these species to the seasonal and year-to-year variability. At Stillwater Cove, seasonal changes are largely the result of changes in growth of the perennnial algae, Pterygophora californica, Cystoseira osmundacea and articulated corallines, which all exhibit maximum cover in summer. These plants are rare at the sites north of Santa Cruz.

The annual sporophytes of Desmarestia ligulata var. ligulata are a consistent, abundant part of the summer flora only at Greyhound Rock (Fig. 3). This alga was uncommon at Stillwater Cove and Sandhill Bluff except during the summer 1978 surveys. At that time, it covered more area than any other alga at Sandhill Bluff. Polyneura latissima also has an annual cycle at the northern sites and also reached its peak abundance at Sandhill Bluff during summer 1978. This alga does not occur at Stillwater cove.

Large sea urchins (Strongylocentrotus franciscanus and S. purpuratus) are extremely rare at Stillwater Cove and small individuals (1-3 cm test diameter) are found only occasionally beneath mats of articulated corallines. Sea urchins were probably abundant prior to the early 1960s (2,24), but sea otter (Enhydra lutris) predation since that time removed

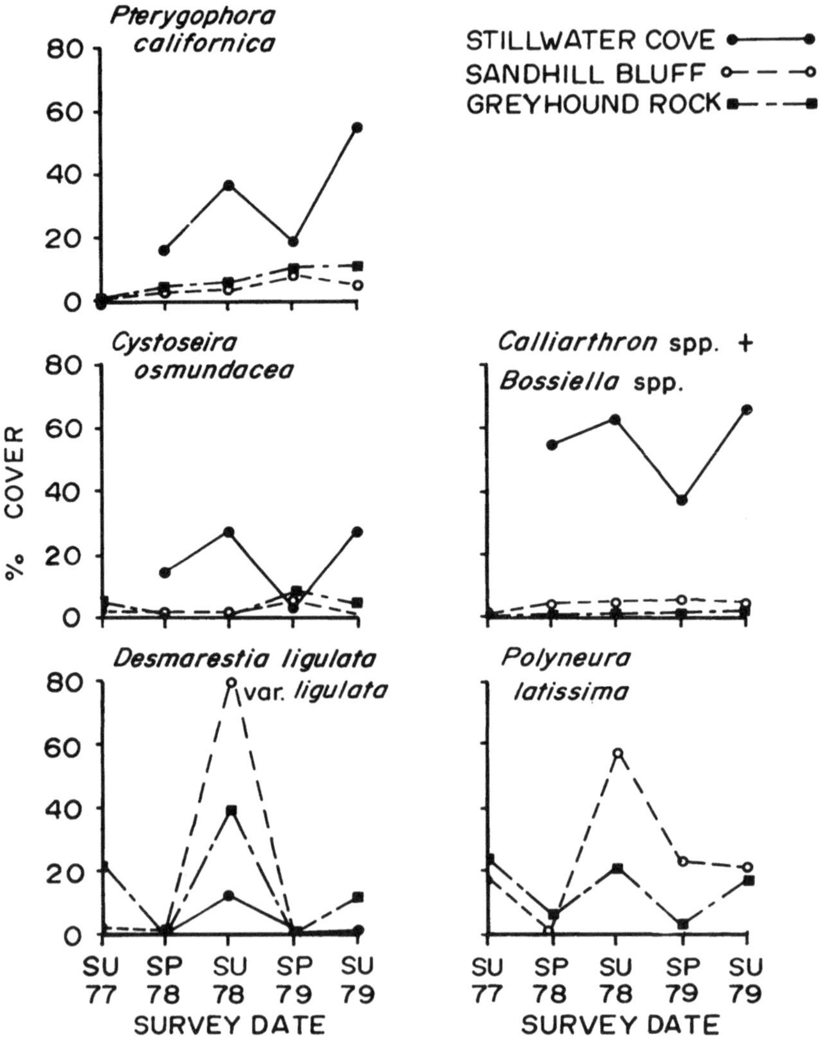

Fig. 3: Cover of common understory algal species at the three survey sites. Data are means (sample sizes in Figure 2); su = summer; sp = spring.

almost all large individuals other than those in inaccessible crevices. Both abalone (Haliotis spp.) and sea urchins are found in areas inhabited by sea otters, if suitable crevices are available (24), but such microhabitats are rare in Stillwater Cove. Grazing snails (Tegula spp.) and crabs (primarily Pugettia producta) are common at Stillwater Cove, but no obvious effects of their grazing have been noted on the algae.

The northern front of the expanding sea otter population in California has not progressed beyond Santa Cruz, and *Strongylocentrotus franciscanus* is common at Sandhill Bluff (Table I). Highest densities occur at the outer edge of the kelp forest, with animals clumped in depressions or along the few available ledges. *S. franciscanus* is very abundant at Greyhound Rock, with highest densities (4-5/m^2) on the walls and valleys between the rock ridges. Encrusting corallines are the most common in these habitats. *Nereocystis luetkeana* and foliose understory algae are most abundant on the ridge tops where sea urchin densities are low. Grazing snails and crabs are extremely rare at both northern sites.

Discussion

The results of the survey, when combined with the physical characteristics of the sites and other information discussed below, suggest that substratum composition, water motion and grazing are of primary direct and indirect importance in regulating the algal associations, but the relative influence of these factors varies between sites. Figure 4 lists the factors for each site and integrates them into regulatory pathways which attempt to explain the observed species composition and abundance.

The substratum characteristics and relatively low water motion at Stillwater Cove are responsible in part for the clear water, resulting in enough light for *Macrocystis pyrifera* to grow in over 30 m of water. Winter storms remove canopy fronds, but usually not entire plants, producing a spring minimum, but allowing rapid recovery to a fall maximum (Fig. 2). *Pterygophora californica* can persist attached to the hard substratum under the low water motion conditions, and occurs in long-lived stands. The changes in cover of this and the other perennial species (Fig. 3) are primarily due to differential blade growth and loss, not recruitment. Smaller perennial understory algae can persist without being removed by storms or buried by sand, and low light produced by the combined canopies of *M. pyrifera* and *P. californica* apparently favors perennial corallines. Rosenthal *et al.* (32) found that *M. pyrifera*

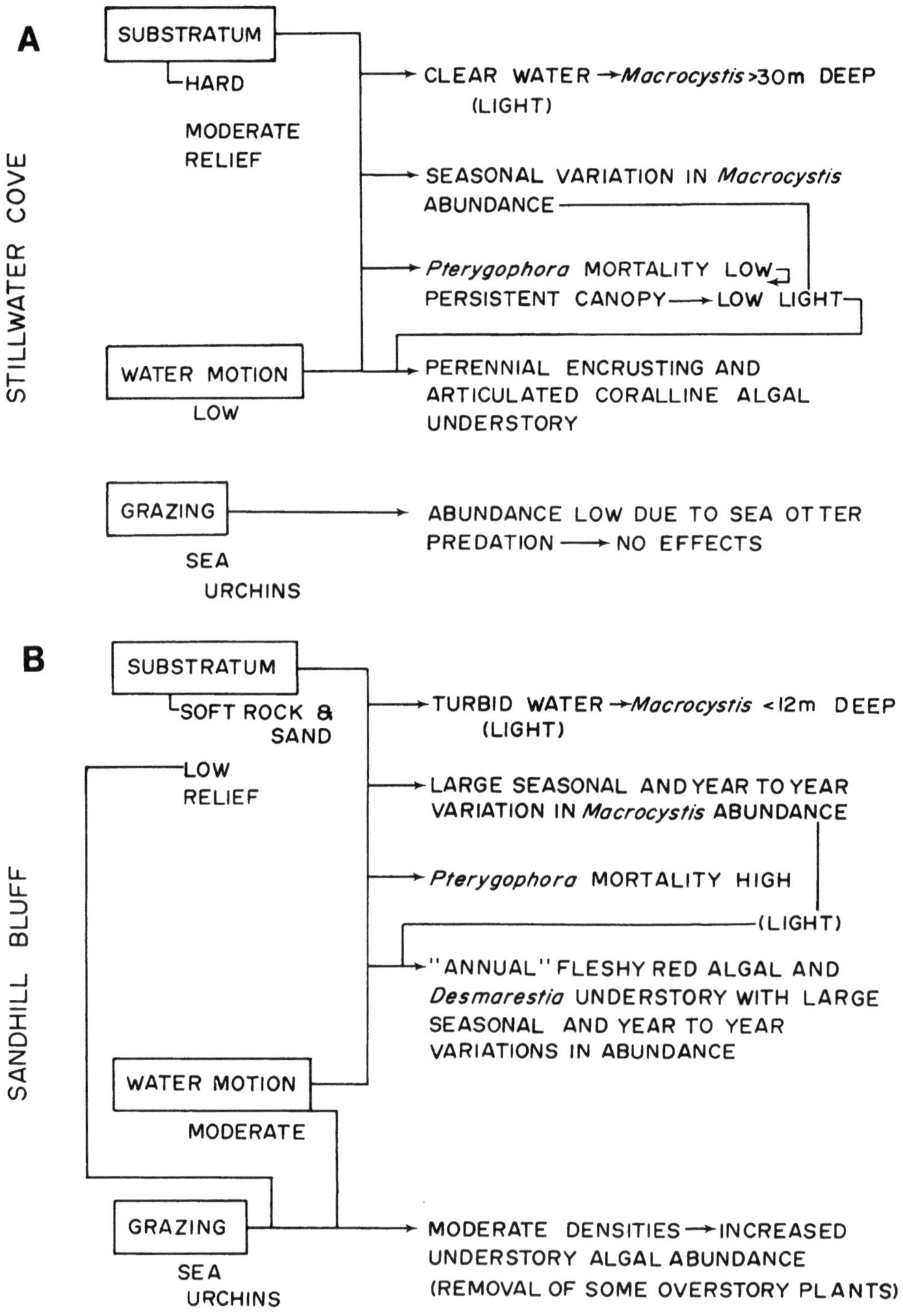
A
STILLWATER COVE
SUBSTRATUM
HARD
MODERATE
RELIEF
CLEAR WATER → *Macrocystis* >30m DEEP
(LIGHT)
SEASONAL VARIATION IN *Macrocystis*
ABUNDANCE
Pterygophora MORTALITY LOW
PERSISTENT CANOPY → LOW LIGHT
WATER MOTION
LOW
PERENNIAL ENCRUSTING AND
ARTICULATED CORALLINE ALGAL
UNDERSTORY
GRAZING
SEA
URCHINS
ABUNDANCE LOW DUE TO SEA OTTER
PREDATION → NO EFFECTS
B
SANDHILL BLUFF
SUBSTRATUM
SOFT ROCK &
SAND
LOW
RELIEF
TURBID WATER → *Macrocystis* <12m DEEP
(LIGHT)
LARGE SEASONAL AND YEAR TO YEAR
VARIATION IN *Macrocystis* ABUNDANCE
Pterygophora MORTALITY HIGH
(LIGHT)
"ANNUAL" FLESHY RED ALGAL AND
Desmarestia UNDERSTORY WITH LARGE
SEASONAL AND YEAR TO YEAR
VARIATIONS IN ABUNDANCE
WATER MOTION
MODERATE
GRAZING
SEA
URCHINS
MODERATE DENSITIES → INCREASED
UNDERSTORY ALGAL ABUNDANCE
(REMOVAL OF SOME OVERSTORY PLANTS)

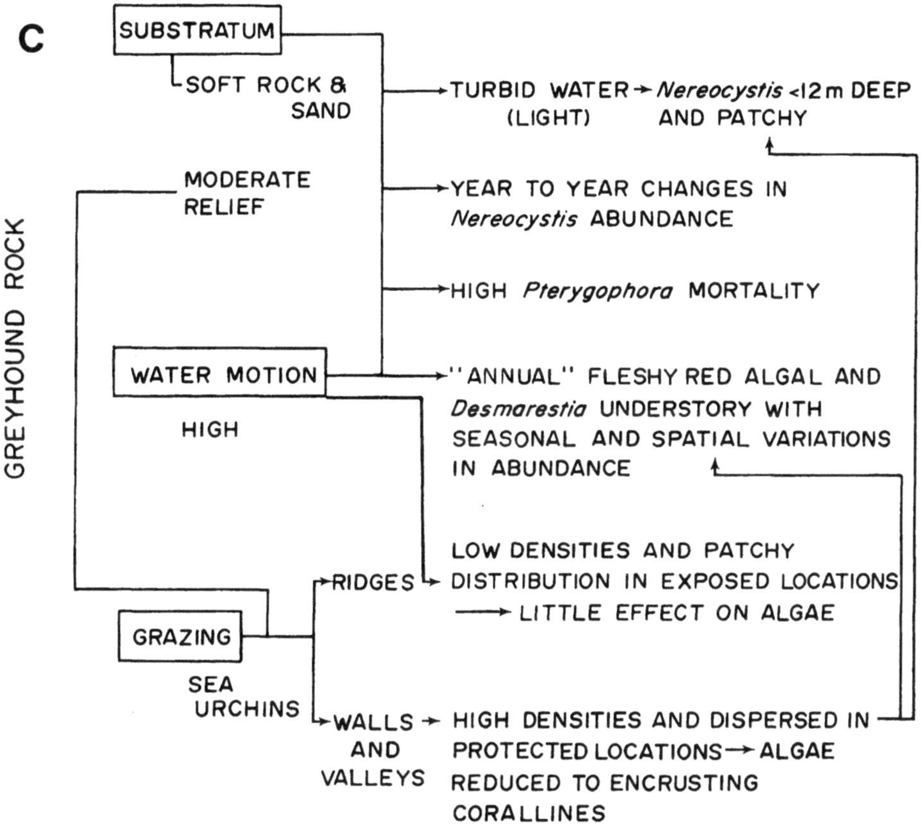

Fig. 4: Factors, factor interactions and regulatory pathways affecting the algal associations in the three survey sites. (A) Stillwater Cove, (B) Sandhill Bluff, (C) Greyhound Rock. Primary factors are shown in boxes, factor interactions by joining lines and regulatory pathways by arrows pointing from cause to effect.

recruitment occurs in patches where groups of adult plants are removed occasionally by storms. This mechanism may be important for M. pyrifera as well as for P. californica in Stillwater Cove.

Other evidence suggests that recruitment and growth of at least brown algae are controlled by competition for light within Stillwater Cove. Desmarestia ligulata var. ligulata was only abundant there during the summer of 1978 (Fig. 3), following a large reduction of the Macrocystis pyrifera canopy (Fig. 2). Moreover, the experimental removal of Pterygophora californica results in a dramatic recruitment of M. pyrifera,

P. californica and D. ligulata var. ligulata (D. Reed, personal communication). The general absence of fleshy red algae may also be related to competition for light, but this has not been investigated. Sea urchins are rare at Stillwater Cove, and other grazers do not appear to have significant effects.

At both Sandhill Bluff and Greyhound Rock, the low light levels produced by resuspended sediments and terrestrial run-off probably reduce successful recruitment and growth of surface canopy kelps below 12 m (Fig. 4). Sea urchins could set this lower limit, but evidence from a physically similar area suggests that plants would not grow in deeper water even if sea urchins were removed (31).

The soft rock and high water motion at Sandhill Bluff interact to cause variations in the Macrocystis pyrifera canopy, either by removing fronds or entire plants. Entire plants were removed during the large swells of winter 1977-78, and the canopy did not recover over the survey area until summer 1979 (Fig. 2). More complete counts of plants and stipes before and after periods of extreme water motion are needed to clarify this interaction. Pterygophora californica does not form stands of large (old) plants at the two northern sites. Perhaps larger plants are more susceptible to removal from the soft rock by water motion.

Our observations suggest that, at both northern sites, the understory algae must be adapted for rapid growth during the periods of abundant light (late spring-summer). In addition, because the combined effects of reduced light, burial, abrasion and removal by storms reduce the chances of winter survival, these plants must be able to persist in a resistent vegetative stage (spores or holdfasts). Articulated corallines and other upright perennial plants may be unsuccessful because their relatively tough thalli are more likely to be completely removed along with portions of the substratum during storms.

The great increase in understory cover at Sandhill Bluff during the summer of 1978 (Fig. 2) was probably a result of increased light caused by the removal of Macrocystis pyrifera during the preceding winter. Other observations indicate competition for space was not important. In the summer of 1977, Desmarestia ligulata var. ligulata was most common just outside the canopy, and there was no reason to assume that space was less available inside the canopy. In addition, as discussed above, D. ligulata var. ligulata increased at Stillwater Cove when M. pyrifera canopy was greatly reduced, and became abundant when the Pterygophora californica canopy was removed. Pearse and Hines (31) also found that the experimental removal of portions of the M. pyrifera canopy at Point Santa Cruz resulted in an increase in understory algae.

The moderate densities of sea urchins at Sandhill Bluff appear to have little negative effect on the vegetation, and may actually increase the abundance and diversity of low-growing algae by removing overstory plants (9). Point Santa Cruz has a Macrocystis pyrifera forest in a habitat similar to that at Sandhill Bluff, but without sea urchins (31). Our surveys at Point Santa Cruz showed the understory algae underwent fluctuations similar to Sandhill Bluff, suggesting that if there are sea urchin grazing effects at the latter site, they are subtle relative to those associated with storms and surface canopy cover.

Nereocystis luetkeana is distributed in patches on the tops of ridges at Greyhound Rock, and our sea urchin removal experiments indicated that it is restricted to these areas by grazing. The increase of this plant after the winter of 1977-78 may have been due to increased substratum availability, a reduction in sea urchins, or some other factor or factors. Understory algae are distributed like N. luetkeana, and this may also be due to sea urchin grazing. The regulatory model (Fig. 4) indicates two possible reasons why understory algal cover did not undergo the large year-to-year variation observed at Sandhill Bluff and other northern Macrocystis pyrifera sites. First, the N. luetkeana canopy is not as dense as that of giant kelp, and reaches its fullest development later in the fall. It has less effect on light and less influence on understory growth. Second, our data show that sea urchins keep large areas of Greyhound Rock relatively free of upright algae at all times of the year, regardless of turbidity, wave action or canopy growth. These areas are thus "stabilized" and act to reduce the overall variation when lumped with areas where algal cover is high and more variable. Sea urchins are probably reduced on the ridge tops at Greyhound Rock because of the high water motion there relative to the walls and valleys. This hypothesis also suggests that sea urchin densities are low at Sandhill Bluff because the relatively flat topography provides few refuges from water motion.

Algal diversity, as measured by the number of species, at the two northern sites is about twice that at Stillwater Cove (Table I). A number of investigators have found that competition can lead to dominance of a few species, and that competitive exclusion often occurs unless some physical or biological disturbance acts to prevent it (see 7 and 8 for reviews). Both large swells and grazing appear to provide such disturbances at Greyhound Rock and Sandhill Bluff, while the lack of grazing and relatively benign physical conditions at Stillwater Cove have apparently favored the establishment of a few long-lived algae which effectively exclude others by shading.

One final question not addressed in figure 4 is: What determines the type of surface canopy kelp at a particular site? The distributions of *Nereocystis luetkeana* and *Macrocystis pyrifera* overlap in central California and historical and recent comparisons at various locations indicate that canopy type is correlated with water motion and sea urchin grazing (31). Our observations also suggest that *N. luetkeana* is more resistant to water motion and more tolerant of sea urchin grazing than *M. pyrifera*; however, *M. pyrifera* may outcompete *N. luetkeana* for light.

Much of the above discussion and many of the proposed interactions are inferred from natural history observations. Quantitative measurements of water motion and light quantity and quality need to be correlated with algal abundance and species composition, and the effects of these factors tested in the field, perhaps with transplant experiments. In addition, further research is necessary to clarify the relationships between the life histories, morphology and growth characteristics of the algae and their respective environments. These data suggest, as proposed by Connell (7) for other communities, that the algal association in kelp forests is an expression of the effects of predation and competition, and that these biological processes are moderated by physical disturbance. Further comparative, long-term surveys and field experiments are needed to test this conceptual framework so that the great differences in kelp forest algal associations can be understood in terms of general ecological processes.

Acknowledgements

I am especially grateful to C. Agegian, R. Cowen, D. Reed, D. Rose and R. VanWagenen for their generous assistance and for sharing their ideas and unpublished results. V. Breda, J. Heine and J. Oliver provided helpful comments on the manuscript. This work was supported by contracts with the United States Marine Mammal Commission and the California Department of Fish and Game.

References

1. Abbott, I., Hollenberg, G.: Marine Algae of California. Stanford University Press, Stanford 1976
2. Andrews, H.: Ecology 26, 24-37 (1945).
3. Barrales, H., Lobban, C.: J. Ecology 63, 657-677 (1975).
4. Broenkow, W., Smethie, W.: Estuarine and Coastal Mar. Sci. 6, 583-603 (1978).
5. Clarke, W., Neushul, M.: In Pollution and Marine Ecology (T. Olsen, F. Burgess, eds.) J. Wiley, New York, pp 29-42 1967
6. Connell, J.: Ann. Rev. Ecol. Syst. 3, 169-192 (1972).
7. Connell, J.: In Ecology and Evolution of Communities (M. Cody, J. Diamond, eds.) Belknap Press, Harvard, pp 460-490 1975
8. Connell, J.: Science 199, 1302-1310 (1978).
9. Cowen, R.: Trophic interactions within a Central California kelp forest. M.S. thesis, California State University, Hayward, 1979
10. Dawson, E., Neushul, M., Wildman, R.: Pac. Nat. 1, 1-80 (1960).
11. Dayton, P.: Fish. Bull. 73, 230-237 (1975).
12. Devinny, J., Kirkwood, P.: Bot. Mar. 17, 100-106 (1974).
13. Devinny, J., Volse, L.: Mar. Biol. 48, 343-348 (1978).
14. Foreman, R.: Helgoländer wiss. Meeresunters. 30, 468-484 (1977).
15. Foster, M.: Proc. Int. Seaweed Symp. 7, 55-60 (1972).
16. Foster, M.: Mar. Biol. 32, 313-329 (1975).
17. Foster, M.: Mar. Biol. 32, 331-342 (1975).

18. Foster, M., Agegian, C., Cowen, R., VanWagenen, R., Rose, D., Hurley, A.: National Technical Information Service (N.T.I.S.) # 293891, Springfield 1979

19. Frye, T.: Publ. Puget Sound Biol. Sta. 2, 65-71 (1918).

20. Greig-Smith, P.: Quantitative Plant Ecology. Butterworths, London, 1957

21. Jackson, G.: Limnol. Oceanogr. 22, 979-995 (1977).

22. Johansen, H., Austin, L.: Can. J. Bot. 48, 125-132 (1970).

23. Leighton, D.L.: Nova Hedwigia 32, 421-453 (1971).

24. Lowry, L., Pearse, J.: Mar. Biol. 23, 213-219 (1973).

25. McLean, J.: Biol. Bull. 122, 95-114 (1962).

26. Neushul, M.: Ecology 48, 83-94 (1967).

27. Neushul, M.: Nova Hedwigia 32, 241-254 (1971).

28. Neushul, M.: In Contributions to the Systematics of Benthic Marine Algae of the North Pacific (I. Abbott, M. Kurogi, eds.), Japanese Society of Phycology, Kobe, pp. 47-74 1972

29. Neushul, M., Foster, M., Coon, D., Woessner, J., Harger, B.: J. Phycol. 12, 397-408 (1976).

30. North, W.J.: Nova Hedwigia 32, 1-97 (1971).

31. Pearse, J., Hines, A.: Mar. Biol. 51, 83-91 (1979).

32. Rosenthal, R., Clarke, W., Dayton, P.: Fish. Bull. 72, 670-684 (1974).

33. ZoBell, C.E.: Nova Hedwigia 32, 269-314 (1971).

DISCUSSION

FOREMAN: Mike, what are your present thoughts about such things as, recruitment, the time and capabilities of certain species to come in and one year graphs, particularly in reference to *Desmarestia vs.* foliose reds, in these northern sites?

FOSTER: To me, the most interesting aspect of these sites is how things like *Polyneura*, *Phycodrys* and other annuals keep coming back year after year, are they coming back from spores or are they coming back from little fragments? Same for *Desmarestia*. It did not occur in any great abundance inside those kelp forests for three years previous to that storm. After the storm it was abundant. So, how did it get there? Is it a persistent gametophyte or is it new settlement from patches on the outside? These sites are also quite different from Stillwater Cove or Coches Prietos Cove, where I did my earlier work, - they are under continual disturbance. At Stillwater Cove, I would probably get results similar to those I got at Coches Prietos Cove. As to diversity, it is very hard to figure out what it means. But, if it is taken to mean number of species, the typical number of species at Stillwater Cove is in the order of ten; typical numbers at these exposed sites, which are unstable, is in the order of 30. So, there may be an inverse relationship between diversity and stability in these two systems, but there is a lot of free space available, if you are ready to colonize, at these northern sites and that could be a more important phenomenon.

FOREMAN: Do you have any evidence of Desmarestia being a preferred urchin food?

FOSTER: The only evidence we have is that during that summer in which Desmarestia was very dense we had another set of sea urchin removal experiments at Sandhill Bluff, which I did not describe. At that time, they had a lot of choices, Desmarestia, foliose red algae, Macrocystis, but, by turning them over to see what was stuck to their jaws, we invariably found Desmarestia, and, although at that time Desmarestia was very abundant, what they had in their stomachs was disproportionate to the abundance in the environment. Now whether that is a true preference or not, I do not know. I should have said, they eat Desmarestia. I was under the impression that it ate their teeth away!

NORTH: Back in the days when the urchins were doing the scraping and I was doing the burning with quick lime, we would go for six miles in the Pt. Loma kelp bed, or what used to be the kelp bed, and there would be almost no algae. The urchins would be completely dominant. We would wipe out the urchins from a section and the first thing that would come in would be diatoms, they would disappear very quickly, then Ectocarpus, that would disappear very quickly, and then it was often Desmarestia. Leighton tried doing some preference experiments in the laboratory with Desmarestia and the urchins in the laboratory would not touch it when other algae were offered. But in the field, it was as you said, they just gobbled it up. When it is there, they will eat it. In the laboratory they will eat something else.

FOSTER: There is a real danger in extrapolating laboratory food preference experiments to the field. There are a number of contradictory examples, not only for sea urchins, but for other kinds of grazing animals.

DRUEHL: We did a Macrocystis harvest here in Bamfield. We looked at the understory and changes in it with complete Macrocystis removal and just canopy removal twice a year. We got the very same story you got with Desmarestia. Our other Laminariales did not respond. Partial harvest -- canopy removal, not plant removal --, gave intermediate results between complete harvest and control.

FOSTER: If you keep that area harvested, do the Laminaria eventually respond to the point that Desmarestia cannot invade that community?

DRUEHL: We kept it harvested for eighteen months and Desmarestia remained dominant.

NORTH: I have seen young plants of Macrocystis come in months after transplants were removed, so that the gametophytes must last a long time. But these were areas where nothing had grown for many years, except Desmarestia was always there, and I think that is a very interesting question. Is it all over the ocean, or is sitting there as a gametophyte or what?

FOREMAN: How do you handle variable quadrat size? As you said you took one to five metre quadrats.

FOSTER: What we are doing is comparing dispersion estimates based on the different quadrat sizes, to see if we can interconvert. If we cannot we will probably express the data in one case as per metre squared, in the other case as per five metre squared and let the reader decide.

FOREMAN: My studies have shown that a large percentage, roughly thirty percent, of all species occuring within the nearshore seaweed beds are sporadic and transient; they may occur one year and never be seen again for five. Some will occur on bare substrate or algal cover, it does not make any difference.

FOSTER: That is interesting because even in these northern areas where, although the species composition fluctuates widely, it still stays the same. In Stillwater Cove, I could tell you at almost any time in the year not only what species you would find there, but what abundance they would be in.

CABOT: Did you ever measure light?

FOSTER: We began to do that last year. We had one of these Licor submersible meters. The problem with all light measurements along this whole coast is that, just from my observations, you might go out one day and it is really turbid and the next day it will be really clear. It is very variable. We can perhaps talk about relative light reduction, certainly in the Pterygophora canopy it is much higher than outside the Pterygophora canopy. But as to absolute values, I do not know what they mean because of the inherent variability from day to day. We also know nothing about how plants respond. Do they respond to two or three days of high light or do they average it? So what we do every time we go out, we take a secchi disc.

ROLES FOR DETRITUS IN COMPLEMENTING PRODUCTIVITY OF COASTAL SYSTEMS

J.R. Sibert
Pacific Biological Station,
Nanaimo, B.C. V9R 5K6 Canada

In this paper I will broadly discuss coastal ecosystem productivity and assess the role of macrophytes in the system. An alternative approach to primary production will be proposed and applied to the analysis of coastal systems.

There are two rather distinct ways of viewing primary productivity. The first is the classical trophodynamic concept articulated first by Elton (2) and later crystallized by Lindeman (4). In a strictly Lindemanian food chain, there are exactly four trophic levels with no place for omnivores. Emphasis is given to the fixation of energy by primary production and its subsequent degradation as material passes up the food chain. Photosynthesis is the only process by which primary production may occur and the two terms have come to be near synonyms. The study of plant growth from this point of view is self-justifying and there is no compulsion to consider the fate of organic matter fixed in the first trophic level.

An alternative point of view is to enlarge the meaning of primary production to "production of nutritious particles for consumers." This definition obviously includes photosynthesis but also admits other processes. Conversion of dissolved organic matter into particles by bacterial activity and transformation of nonnutritious detrital particles into nutritious ones by microbial conditioning may also be included. This point of view requires consideration of the fate of organic matter at the first trophic level and is directly relevant to fisheries studies.

Synthetic and Degradative Processes in Marine Macrophytes

Table I: Relative importance of imported carbon for two coastal ecosystems of different size. Figures recalculated from various sources as shown.

	System	
	Nanaimo Delta	Strait of Georgia
Surface area (km^2)	6.5[a]	6900[b]
Carbon source ($gC\ m^{-2}\ yr^{-1}$)		
Fluvial DOC	2000[a]	130[a]
Phytoplankton	7.5[a]	120[e]
Macroalgae	0.9-7.5[d,e]	2.7-6.7[f]

[a]Naiman and Sibert (6) give the average DOC concentration in the Nanaimo River as 6.44 mg Cl^{-1}. This estimate is multiplied by the volume of fresh water entering the Strait ($145 \times 10^9\ m^3$) and divided by the surface area.
[b](15)
[c](8)
[d]Foreman, R.E., personal communication.
[e](7)
[f]Foreman estimates total algal biomass in Strait of Georgia at 27-67 $\times 10^9$ grams dry weight. This estimate was increased by 2.3 to correct for mortality and exudation and a carbon content of 30% was assumed (personal communication).

Applicability of this enlarged productivity definition depends on both the boundaries of the system and its size or scale (11). In a very large, or global ecosystem, or one with boundaries closed to material imports, it is obvious that the only process capable of introducing energy is photosynthesis. In much smaller systems such as river deltas or segments of the coastal zone, large amounts of material are imported from rivers and adjacent areas by longshore transport. Some effects of scale are illustrated in Table I, for two coastal systems in British Columbia. The estimates in the table were compiled from various sources and represent averages for the entire system, that is the total input of material divided by the surface area of the system. The large amount of dissolved organic carbon (DOC) introduced by the Nanaimo River overwhelms all other sources of carbon, a conclusion previously reached by Seki et al. (9). This effect is diminished in the Strait of Georgia by the much greater surface area. I do not mean to suggest that size is the only important difference between the Nanaimo Delta and the Strait of Georgia. The point is that the relative importance of imports depends on size of the system.

The calculations in Table I assume that the entire production of seaweeds along the rock margins of the Strait of Georgia is uniformly distributed throughout the Strait. While there are few data on this subject, Smith (12) suggests that most kelp detritus remains in the vicinity of the beds where it was produced. The capacity to retain materials may be a very important feature of open systems and probably contributes to the maintenance of high productivity of coastal ecosystems (14,16).

Detritus may also provide a productive basis for coastal ecosystems at times of the year when photosynthesis is low. Figure 1 shows abundance and productivity cycles for several important components of estuarine systems in the Strait of Georgia and elsewhere. Dissolved organic carbon (DOC) is relatively high and constant throughout the year and has a maximum in April. Phytoplankton productivity is low during most of the year with an early summer maximum. Sedge biomass is low in winter and shows a prolonged late summer maximum. Benthic microalgae productivity on sand flats is lowest in autumn and early winter with a gradual increase beginning in January and reaching a maximum in mid summer. Benthic macroalgae, primarily Ulothrix flacca and Pylaiella littoralis growing in the marsh, attain maximum biomass in autumn and mid-winter.

Secondary production, illustrated by meiofaunal abundance, seems to have two peaks, one in spring and a second later in the summer. The summer maximum may be associated with blooms of benthic microalgae or with detritus derived from other summer productivity sources. The basis for the spring secondary production is less obvious. Finally, juveniles of two commercially important species of fish, Pacific herring and chum salmon, attain their maximum numbers in coastal nursery areas in early spring. Since both of these species suffer large juvenile mortality, the basis of their productivity is of some practical interest. Data on herring feeding and growth in the coastal zone are scarce but chum fry in the Nanaimo estuary are known to depend heavily on a single species of harpacticoid copepod which also attains its maximum biomass in early

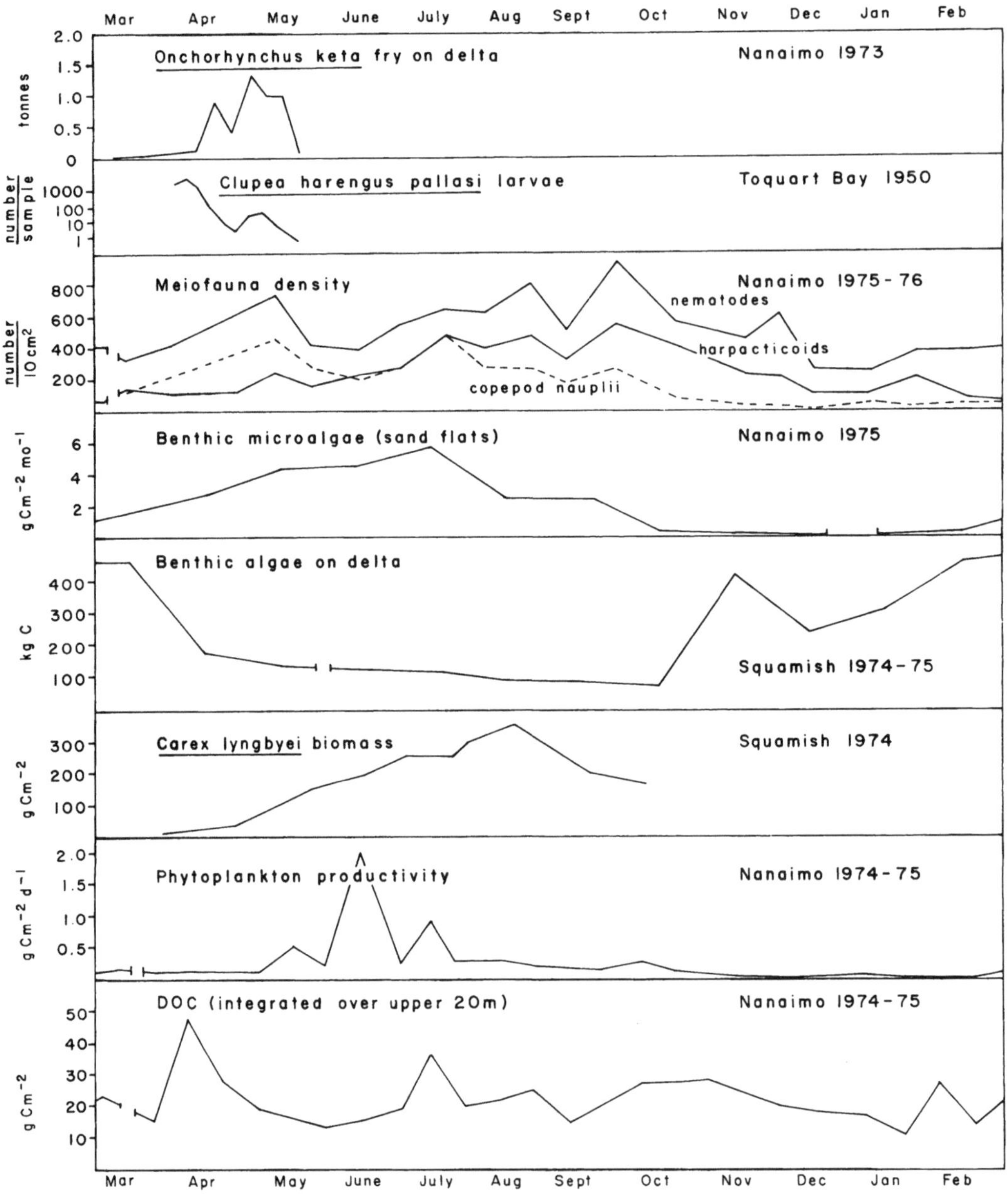

Fig. 1: Abundance and productivity cycle for important components of some estuarine systems. Data taken from various sources: from top to bottom the references are 3,13,10,7, Pomeroy, W.M., personal communication, 5,1,1.

spring (3,10). The relative timing of these events is such that the productivity of the juvenile fish is more likely to depend on detritus and winter algal production than on the customary summer productivity cycles. Degradation of marine macrophytes is an obvious source of detritus to drive secondary production in the coastal zone.

In summary, the concept of primary productivity may be expanded to include all processes which generate nutritious particles for consumers. This definition is particularly important to the coastal zone where heterotrophic processing of imported detritus may greatly augment photosynthetic processes in the production of nutritious particles. From this point of view, detritus has two functions: 1) to increase total system productivity in the summer; 2) to provide a productive base during times of the year when photosynthesis is low. This latter function may particularly be important for survival of juveniles of commercial fish which occupy the coastal zone prior to the major bursts of photosynthetic productivity.

References

1. Brown, T.J., Stephens, K.V.: Fish. Mar. Serv. Data Rep. 20, 146 pp. (1976).
2. Elton, C.: Animal Ecology. Sidgwick and Jackson, London, 1927
3. Healey, M.C.: J. Fish. Res. Board Canada 36, 488-496 (1979).
4. Lindeman, R.L.: Ecology 23, 399-418 (1942).
5. Levings, C.D., Moody, A.I.: Fish. Mar. Serv. Res. Dev. Tech. Rep. 606, 51 pp.
6. Naiman, R.J., Sibert, J.R.: Limnol. Oceanogr. 23, 1183-1193 (1978).
7. Naiman, R.J., Sibert, J.R.: J. Fish. Res. Board Canada 36, 504-520 (1979).
8. Parsons, T.R., LeBrasseur, R.J., Barraclough, W.E.: J. Fish. Res. Board Canada 27, 1251-1264 (1970).
9. Seki, H., Stephens, K.V., Parsons, T.R.: Arch. Hydrobiol. 66, 37-47 (1969).
10. Sibert, J.R.: J. Fish. Res. Board Canada 36, 497-503 (1979).
11. Sibert, J.R., Naiman, R.J.: In Marine Benthic Dynamics, (K.R. Tenore, B.C. Coull, eds.), Univ. South Carolina Press, pp. 311-323 1980

12. Smith, B.D.: A qualitative and quantitative assessment of seaweed decomposition in the Strait of Georgia. M.Sc. thesis, Department of Botany, University of British Columbia. 168 pp. 1979

13. Stevenson, J.C.: J. Fish. Res. Board Canada 19, 735-810 (1962).

14. Valiela, I., Teal, J.M., Volkmann, S., Shafer, D., Carpenter, E.J.: Limnol. Oceanogr. 23, 798-812 (1978).

15. Waldichuk, M.: J. Fish. Res. Board Canada 14, 321-486 (1957).

16. Woodwell, G.M., Whitney, D.E., Hall, C.A.S., Houghton, R.A.: Limnol. Oceanogr. 22, 833-838 (1977).

DISCUSSION

DRUEHL: Could you test possible contributions of run off from rivers in an area where you have different seasons of run off, but maybe the same types of early season fish stock?

SIBERT: Yes, I think that would be an interesting thing to do. We could easily compare Nanaimo, which has a fall-winter run off, with a place like Squamish which has glacial run off.

COON: Do you have any information on effects of logging on stream run off and its subsequent effect on estuaries?

SIBERT: Well, there is a lot of work at Carnation Creek on the detrital production in the stream in relation to fresheting and logging events, but they are not looking at the estuary.

DUNCAN: Even though the greatest amount of carbon is coming down from a river, the contributions from the other carbon producers in the estuary may be significant at some times.

SIBERT: Well, the point is that you get carbon coming down the river, say at Nanaimo, at a time of the year when other sources of production are very low, photosynthesis is low, so it has the effect of smoothing out the curve for total carbon. The problem is we do not know if the dissolved carbon coming down the river in the fall is susceptible to microbial attack, and also, we do not know how much of it is retained. It could just go gushing right through.

CABOT: Did you find any utilization of *Pylaiella*?

SIBERT: That is apparently utilized by gammarids. I do not think it is a favorite, they prefer something else.

DEGRADATION OF THE KELPS MACROCYSTIS INTEGRIFOLIA AND NEREOCYSTIS LUETKEANA IN BRITISH COLUMBIA COASTAL WATERS[1]

L.J. Albright, J. Chocair, K. Masuda and M. Valdés
Department of Biological Sciences, Simon Fraser University
Burnaby, B.C. V5A 1S6, Canada

Introduction

A significant proportion of the sub-tidal rocky shoreline of coastal British Columbia is colonized by relatively extensive beds of macroalgae, including those of Macrocystis integrifolia and Nereocystis luetkeana (5,6,21). Where present, these large kelps are believed to significantly contribute to marine food webs by several means. These include: 1) excretion of dissolved organic matter, which is subsequently metabolized by free-living and attached heterotrophic bacteria (10) and animals (7) and 2) sloughing of kelp tissues which are rapidly colonized and degraded by microorganisms. Colonized detritus particles such as these may in turn be a significant food source for members of higher trophic levels of coastal marine food chains (22).

This investigation was undertaken to evaluate the importance of M. integrifolia and N. luetkeana to the microbial and animal portion of the marine food web in and adjacent to kelp beds by 1) determining in situ rates and extents of kelp degradation, 2) observing the succession of microbial species and numbers associated with kelp detritus at various stages of degradation and 3) evaluating the nutritive quality of decomposing kelp tissues.

[1] Portions of this paper have been previously published in Naturaliste Canadien 107: 3-10 (1980). We thank the publishers of this journal for permission to copy parts of that manuscript in this work.

Synthetic and Degradative Processes in Marine Macrophytes

Materials and Methods

Collection and incubation of kelp. *M. integrifolia* was collected (5 August 1977) and incubated at a sheltered bay at Dixon Island (48° 51' 16" N; 125° 6' 36" W) approximately 2 km from the community of Bamfield, British Columbia. *N. luetkeana* was harvested and placed for degradation (28 August 1976) at Brockton Point (49° 17' 50" N; 123° 6' 54" W) within the harbour limits of Vancouver, British Columbia.

Immediately after harvest, 100 g portions (wet weight) of laminae and stipes of *M. integrifolia* and laminae and pneumatocysts of *N. luetkeana* each were placed in litter bags of 1.5 mm mesh size. These bags were incubated both in the water column at 10 m depth, approximately 50 m from shore, and on the substratum of the intertidal region at each location. Litter bags were removed at 5-8 day intervals and their contents were assayed as described below.

In situ degradation analyses. The degrading kelp retained in each litter bag, was weighed at specified time intervals. Each wet weight determination was in duplicate and the values were averaged.

Scanning electron microscopy. Kelp samples were placed in 2.5% glutaraldehyde (pH 7.0) immediately after removal from the marine environment. Each sample was then critical point dried and very lightly sprayed with a 1% solution of Kodak Photo-flo 200. Following these treatments, samples were examined using an ETEC autoscan scanning electron microscope.

Enumeration of viable bacteria. Viable bacteria were determined by the method described by Laycock (13). In essence, kelp tissues were cycle-comminuted using a cooled Waring blender and appropriate dilutions of each slurry were spread and plate counted on 2216E marine agar (Difco). Plates were incubated for 15 days at 15 C before being counted. The viable heterotrophic bacteria/g of the contents of each litter bag were then calculated.

Total bacterial numbers were determined by fluorescent microscopy using the acridine orange direct count (AODC) technique as described by Hobbie et al. (12).

ATP analyses. The material cycle-blended as described above was analyzed for ATP by the method of Bancroft et al. (1).

Chlorophyll a analyses. Chlorophyll a was analyzed as described by Strickland and Parsons (25).

Crude fibre, carbon, nitrogen and ash determinations. Crude fibre analyses were performed by the method described in Strickland and Parsons (25, pp. 235-236). Total carbon and nitrogen were assayed with the use of a CHN analyzer (Beckman), using the technique described by Menzel and Vaccaro (16).

Kelp samples of known wet weight, previously blotted to remove sea water, were heated to 95 C for 3 days till constant weight, to determine dry weights. These dried tissues were then fired at 500 C for 2 h and the residues weighed. These values were used to calculate ash weight/g dry weight kelp.

Results and Discussion

When 80 g portions of N. luetkeana laminae are cycle-comminuted for 30 s, placed in 920 ml of sea water in a 2 l Erlenmeyer flask and incubated at 15 C in the dark, the numbers of bacteria increase dramatically (Fig. 1). Within 4 days the numbers reach 10^{10}/ml of kelp suspension when assayed by the plate count technique and approximately 5×10^{11}/ml by the AODC technique. Assuming each bacterial cell has a mass of 1×10^{-13} g then the 80 g of laminae produced 1 g (plate count), or 50 g (AODC technique) of bacterial biomass for bacterial production efficiencies from kelp of approximately 1 or 62%, respectively. Since direct counts are a better assay of bacterial numbers than plate counts, the latter figure is probably the more accurate of the two. It also compares favourably with the approximately 41% efficiency for conversion of Laminaria detritus to bacterial biomass noted by Robinson et al. (19). These rapid and efficient conversions of kelp to bacterial biomass illustrate the potential for degrading kelp to influence bacterial growth rates and activities in coastal ecosystems.

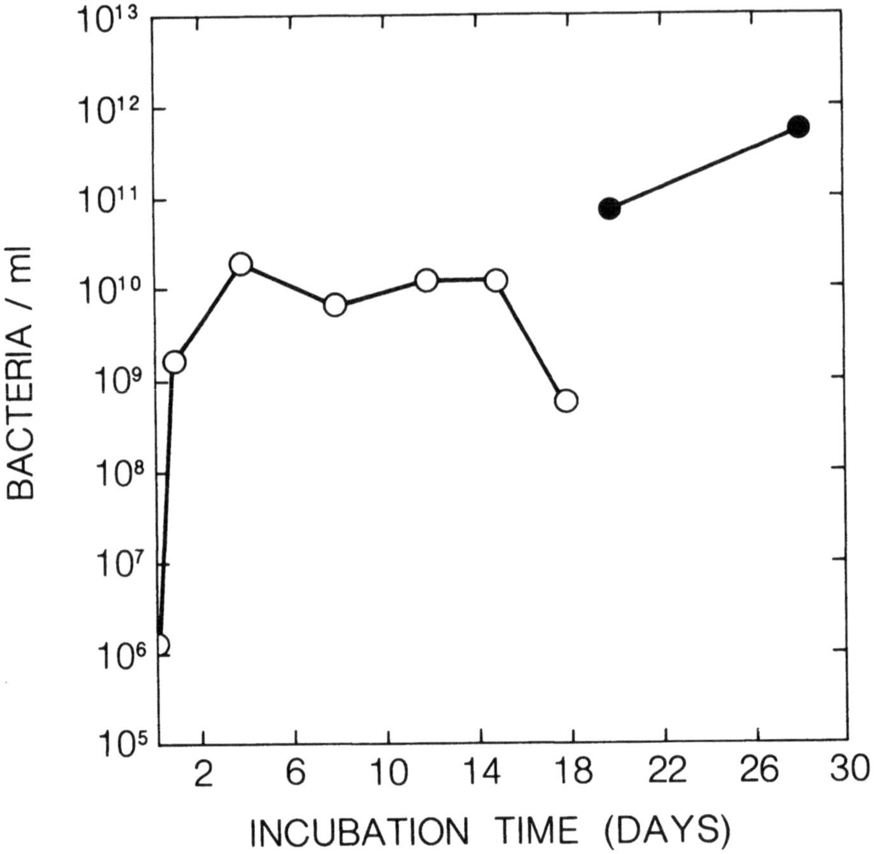

Fig. 1: Growth of bacteria in a comminuted kelp-seawater system when assayed by the plate count (o) and direct count techniques (●).

In late summer and early fall (August-October), _in situ_ degradation of tissues of _M. integrifolia_ and _N. luetkeana_ within litter bags to mineralized components (e.g. carbon dioxide and water), dissolved organic matter and particulate organic matter (1.5 mm or less in size) proceeds relatively rapidly in both the sub- and intertidal portions of the marine environment of southern coastal British Columbia. Times of degradation to residues of approximately 1.5 mm or less varied from 5 to 8 weeks (Fig. 2). Other investigators have reported that within an 8 week period the salt marsh plants _Juncus_, _Distichlis_, _Spartina_ and _Salicornia_ spp. decayed to approximately 70, 50, 45 and 10% of their original weights respectively, when contained within 2.5 mm mesh size litter bags and placed in a marine marsh (17). Morrison et al. (15) found that leaf samples of the oak Quercus virginiana and the pine Pinus eliottii lost approximately 15 and 40% of their original weights 6 weeks after being placed in a marine estuary. Thus, the rates of kelp degradation in coastal British Columbia waters are probably among the greatest for plant material decaying in the temperate marine environment. Indeed, these degradation rates would probably have been greater if the kelp tissues were not contained within litter bags. Physical stresses due to wave action and water current movements would probably increase the degradation rates (8,18).

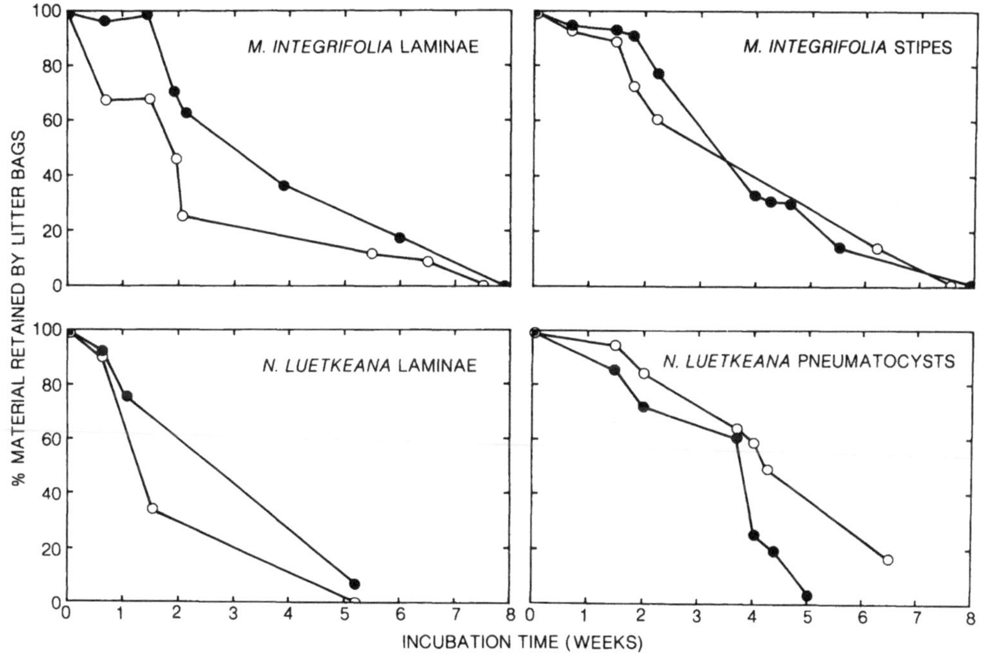

Fig. 2: Rates of degradation of M. integrifolia (laminae and stipes) and N. luetkeana (laminae and pneumatocysts) incubated at Dixon Island and Brockton Point, respectively. The symbols (o) and (•) denote samples incubated inter- and subtidally, respectively.

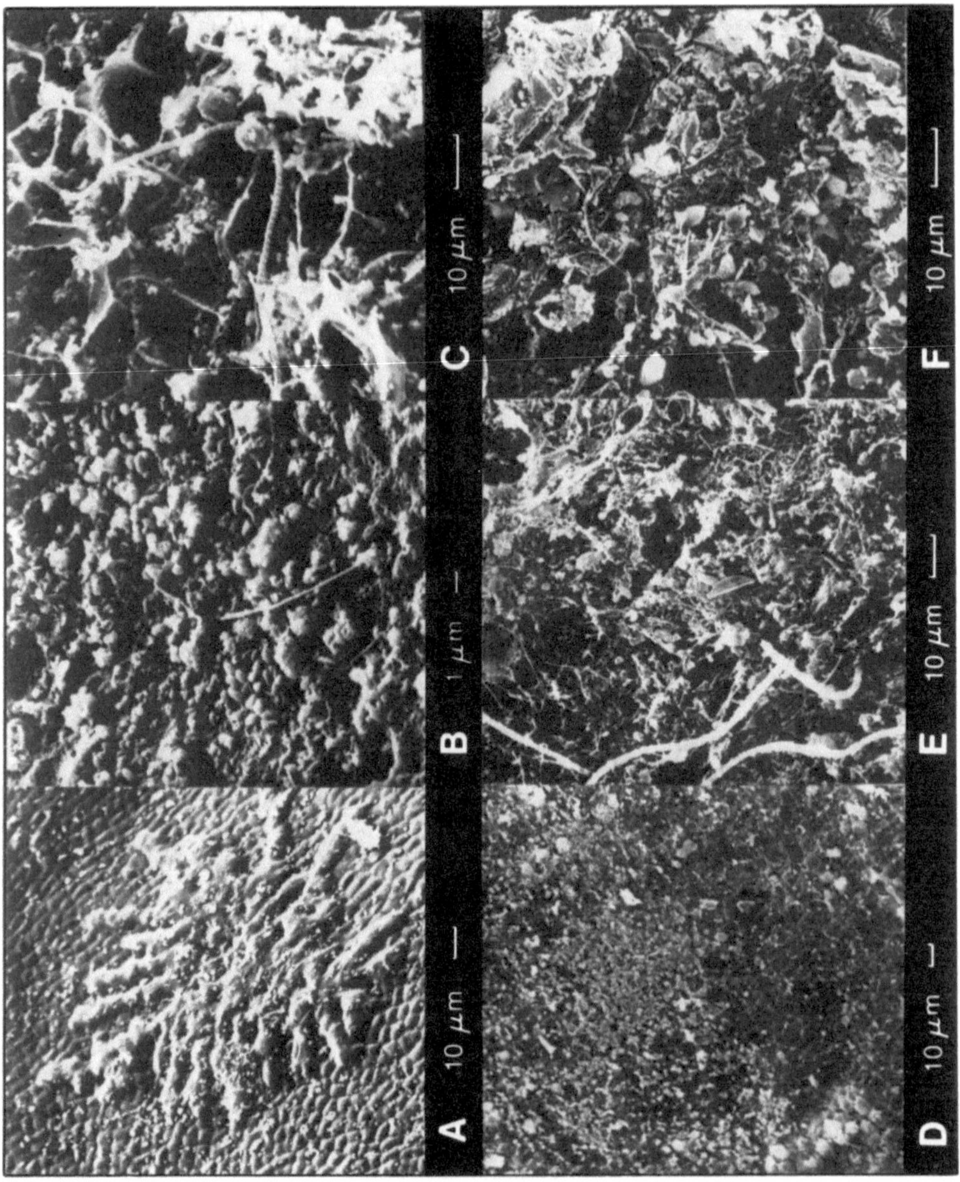
A 10 μm
B 1 μm
C 10 μm
D 10 μm
E 10 μm
F 10 μm

Fig. 3: Scanning electron micrographs of colonization of N. luetkeana laminae and pneumatocysts during incubation for various lengths of time in litter bags suspended in the water column. (A) lamina surface, day 0; (B) lamina surface, day 3; (C) lamina surface, day 13; (D) pneumatocyst surface, day 0; (E) pneumatocyst surface, day 4; (F) pneumatocyst surface, day 22.

The microbial populations, as observed with the scanning electron microscope, associated with healthy and decaying N. luetkeana tissues, are displayed in Fig. 3. In general, we noted that the surfaces of healthy tissues of this kelp were sparsely populated by pennate diatoms, yeast-like cells, single or diploid rods, or more commonly, microcolonies of end-attached bacteria which appeared to be very similar to the epiphytic bacterium Leucothrix mucor described by Bland and Brock (2). The presence of small detritus particles on healthy tissues was common.

Within 3 days of placement of healthy kelp tissues in litter bags in the marine environment significant quantities of both detritus particles and bacteria appeared on the kelp (Fig. 3). By 5 days, large numbers of pennate diatoms also settled. As degradation proceeded other microbial types such as centric diatoms, yeast-like cells and protozoa colonized the kelp tissues as well. By approximately day 14 a complex matrix of detritus, bacteria, algae, fungi and protozoa coated the decaying kelp (Fig. 3). The approximate time order of settlement and colonization was detritus, bacteria, followed by algae, fungi, protozoa.

Cundell and Mitchell (4) noted that the sequence of microorganisms colonizing a wood surface immersed in sea water was: bacteria followed by detritus, pennate diatoms, stalked diatoms, protozoa, macroalgae, barnacles and isopods, in that order. Morrison et al. (15) also noted that the initial colonizers of both oak and pine leaves placed in the marine environment were bacteria followed by a variety of eucaryotes, both algae and fungi. Other workers have found similar patterns of microbial colonization on materials placed in the marine environment (9,23,26).

The rapid colonization of decaying kelp tissues by bacteria, as observed by scanning electron microscopy, was also noted upon assay of total viable heterotrophic bacterial numbers (colony forming units - CFU) using the spread plate technique. These numbers increased from approximately 10^5/g of healthy kelp tissue, harvested in late summer, to approximately 10^{11}/g of decaying tissue within 1.5 to 5 weeks of placing the material in litter bags (Fig. 4). In all samples studied, the kelp tissues placed in the intertidal environment more quickly attained greater numbers of viable heterotrophic bacteria than those placed at 10 m depth in the water column (Fig. 4). Since the bacterial contents of surface sediments usually exceed those of overlying waters by at least two orders of magnitude and the litter bags (with contents) rested on sediments at low tides, the greater initial colonization rates of the intertidal incubated samples are understandable. However, the reason for the greater (approximately ten-fold) extent of bacterial colonization of intertidal incubated samples is not known with certainty. Two possible explanations for this enhancement are: 1) nutrients from sediments enhanced the growth of bacteria colonizing the intertidally decaying kelp; and 2) many of the bacteria were adventitious species washed onto the intertidally incubated kelp from sediments. The numbers of viable heterotrophic bacteria/g of healthy kelp tissues are approximately the same as those reported by Laycock (13) for _Laminaria longicruris_ and Chan and McManus (3) for _Polysiphonia lanosa_ and _Ascophyllum nodosum_.

On a per unit wet weight basis, the ATP content of healthy _M. integrifolia_ laminae is very much greater than that of stipes (Fig. 5). However, upon placing these tissues in litter bags within the marine environment, ATP levels of both tissues approached minimal values within approximately one week. If ATP content is an index of kelp tissue viability it would appear that the kelp fronds and laminae were "dead" within one week of placement within litter bags. The subsequent gradual increases in ATP content of both decaying laminae and stipes were probably due to microbial colonization and growth on these tissues. Scanning electron microscopy indicated microbial colonization within 3 days of placement in the sea; however, the microbial biomasses were probably not sufficiently large within the first week of incubation to greatly influence total ATP content of the decaying matter. Relatively large increases in ATP levels occurred only after approximately 3 weeks incubation _in situ_. This was probably due to microbial attachment and growth with a relatively large contribution from the heterotrophic

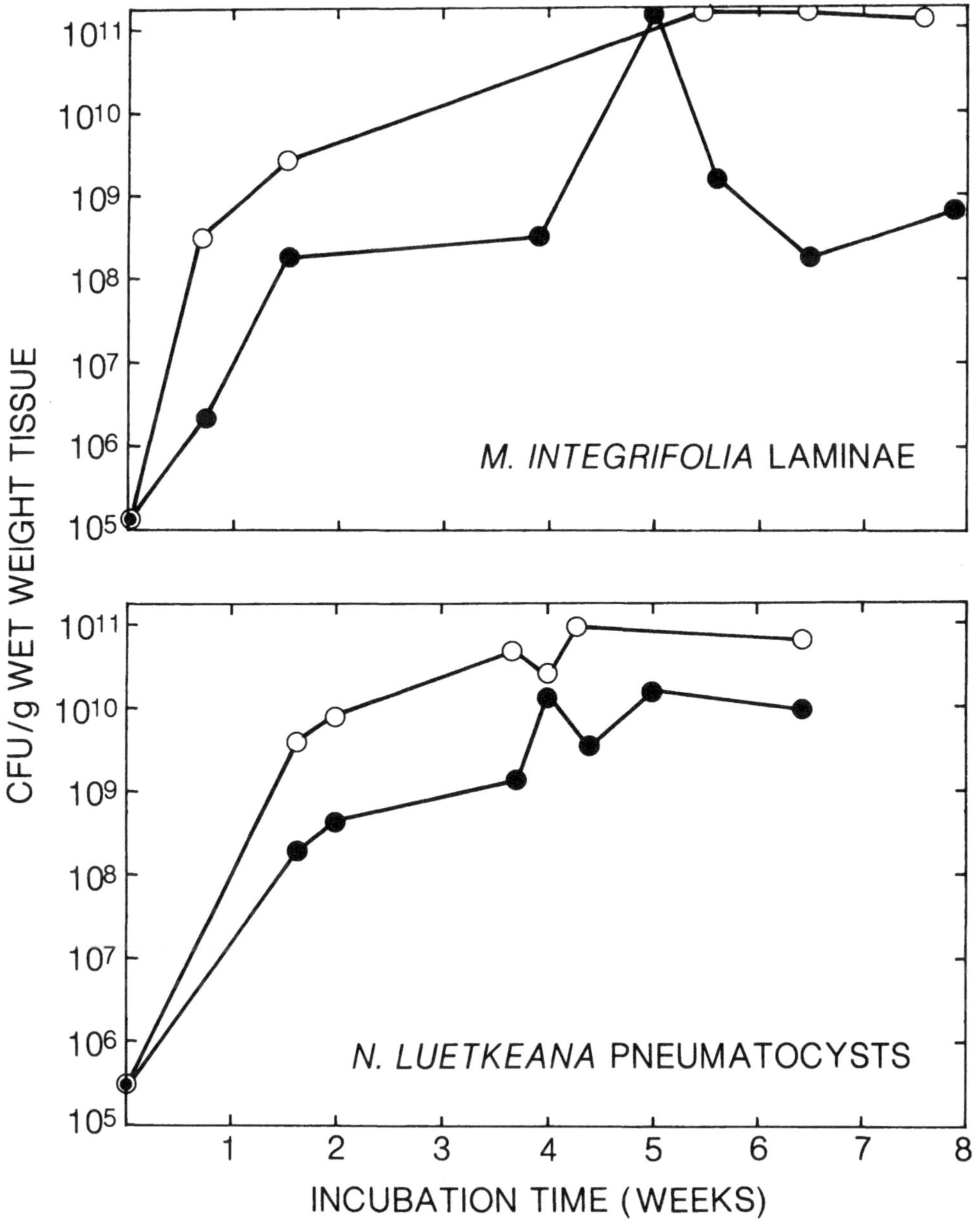

Fig. 4: Viable heterotrophic bacterial counts of decaying M. integrifolia laminae and N. luetkeana pneumatocysts. (o) and (●) denote samples incubated inter- and subtidally, respectively. Rates and extents of bacterial colonization of M. integrifolia stipes and N. luetkeana laminae were similar.

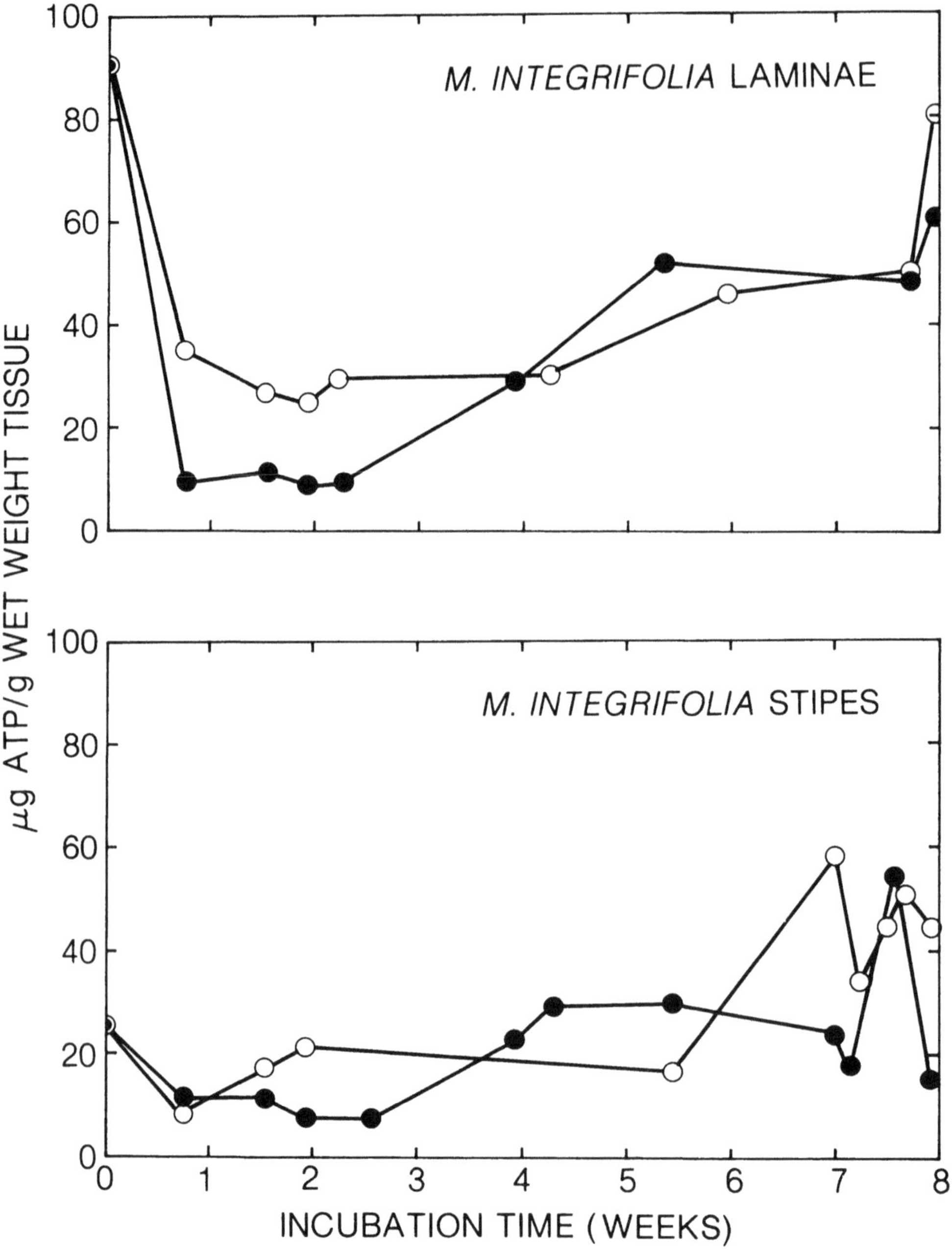

Fig. 5: ATP content of decaying M. integrifolia laminae and stipes incubated inter- (o) and subtidally (●).

bacteria (see Figs. 3 and 4). Since heterotrophic bacterial numbers reached maximal values in 4-5 weeks, increased ATP concentrations observed at the end of 3 weeks were probably due to increases in numbers of other microorganisms, e.g. algae, fungi and protozoa (see Fig. 3). On the basis of ATP data alone, however, one cannot determine the relative contribution of each type of microbe to the total microflora at any one time.

Chlorophyll a content of M. integrifolia laminae is much greater than that of stipes (Fig. 6), probably because the laminae are the major area for photosynthesis of this marine plant. When both laminae and stipes, contained in litter bags, were placed in the inter- and subtidal marine environment the chlorophyll a content decreased, probably due to microbial degradation. However, within 1-3 weeks chlorophyll a levels increased. This is indirect evidence that the majority of the initial colonies were probably heterotrophic microorganisms with a greater proportion of the later-colonizing ones being microalgae.

Both healthy and decaying tissues of M. integrifolia and N. luetkeana are believed to be food sources for microorganisms and animals (see Figs. 3 and 4; [14], and L. Druehl, personal communication). Therefore, to assess the relative value of M. integrifolia as an animal food, analyses of this kelp, both healthy and at various stages of decomposition, were done for crude fibre, percent carbon and nitrogen as well as ash weight.

Along with the loss of organic matter in decaying macrophytes in the marine environment, there are apparent changes in nutritive content. Crude fibre content of M. integrifolia laminae marginally decreased with time, whereas that of stipes remained relatively constant in both inter- and subtidal environments (Fig. 7). These results are in general agreement with those of other investigators who have noted that crude fibre content of Spartina derived detritus remained relatively constant (17) and that of decaying red mangrove leaves showed a slight increase (11).

The carbon content of both healthy laminae and stipes of M. integrifolia was approximately 30%, whereas the nitrogen content of laminae was approximately twice that of stipes (Table I). As degradation of these

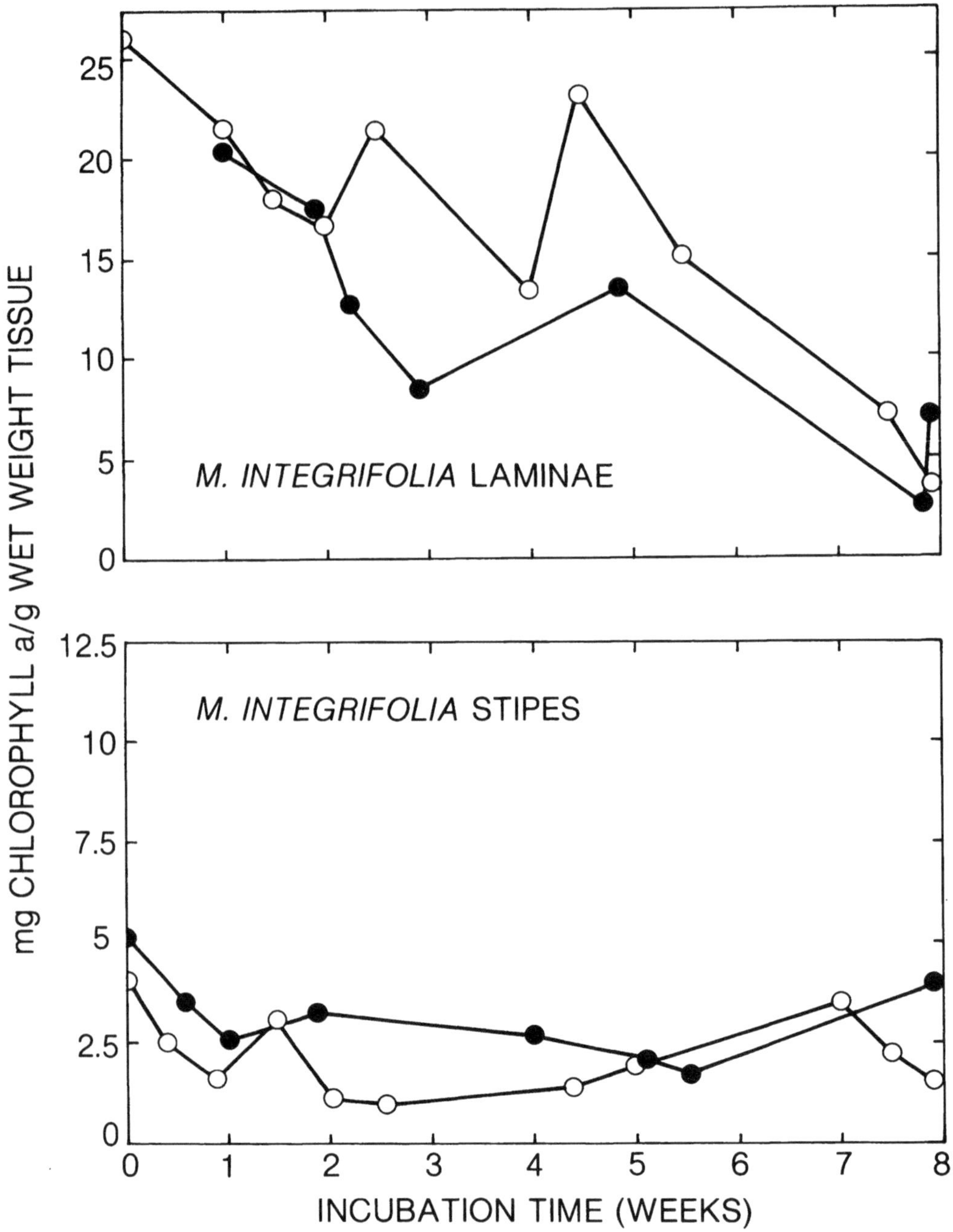

Fig. 6: Chlorophyll a content of decaying *M. integrifolia* laminae and stipes incubated inter- (o) and subtidally (●).

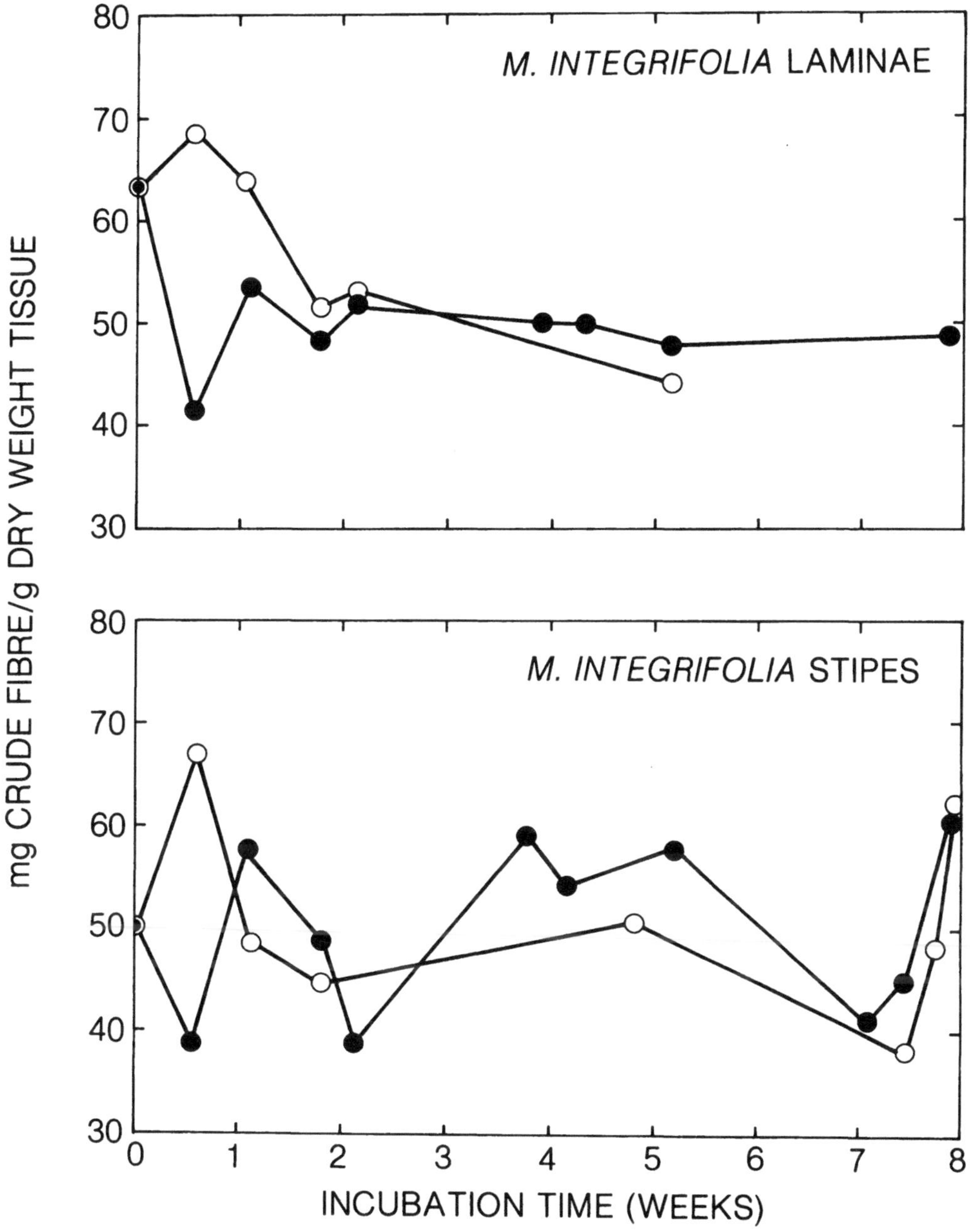

Fig. 7: Crude fibre content of decaying M. integrifolia laminae and stipes incubated inter- (o) and subtidally (●).

tissues within litter bags proceeded, the carbon content decreased to approximately 24-26%, whereas that of nitrogen increased (see Table I).

Table I: Total carbon and nitrogen content of M. integrifolia tissues degraded in litter bags subtidally[a] at 10 m depth in the water column. Material was harvested in late summer.

Incubation Time (weeks)	Laminae %C	%N	C/N	Stipes %C	%N	C/N
0	30.0	2.1	14.3	29.5	1.0	29.5
1	29.1	2.1	13.9	24.6	0.9	27.3
2	29.0	2.3	12.6	25.8	0.9	28.7
3	29.8	2.2	13.6	25.8	1.2	21.5
4	29.8	2.2	13.6	27.2	1.8	15.1
5	26.9	2.5	10.8	30.1	1.6	18.8
6	26.5	2.5	10.6	28.0	1.3	21.5
7	26	2.5	10.4	24.1	1.5	16.1
7.8	26.1	2.5	10.4	-	-	-

[a] Data for samples incubated intertidally showed similar patterns.

In late summer, ash content of healthy M. integrifolia stipes, on a dry weight basis, was somewhat greater than that of the laminae. However, upon degradation within litter bags, ash content of the laminae increased, whereas that of the stipes remained relatively constant (Fig. 8). The samples incubated intertidally tended to have a greater ash content than the ones incubated in the subtidal water column. Since the intertidal samples were associated with sediment, they may more readily have accumulated some inorganic elements.

It has been suggested (20) that the required C:N ratio for marine animal food, as detritus should be at least 17:1. Diets with greater C:N ratios may be deficient in nitrogen compounds, particularly protein. Our results indicate that healthy M. integrifolia laminae may be of nutritionally better quality in this sense than stipes. The laminae showed C:N ratios of approximately 14.3:1 vs. 29.5:1, for stipes in material harvested in late summer (Table I). Moreover, the C:N ratios of both tissues

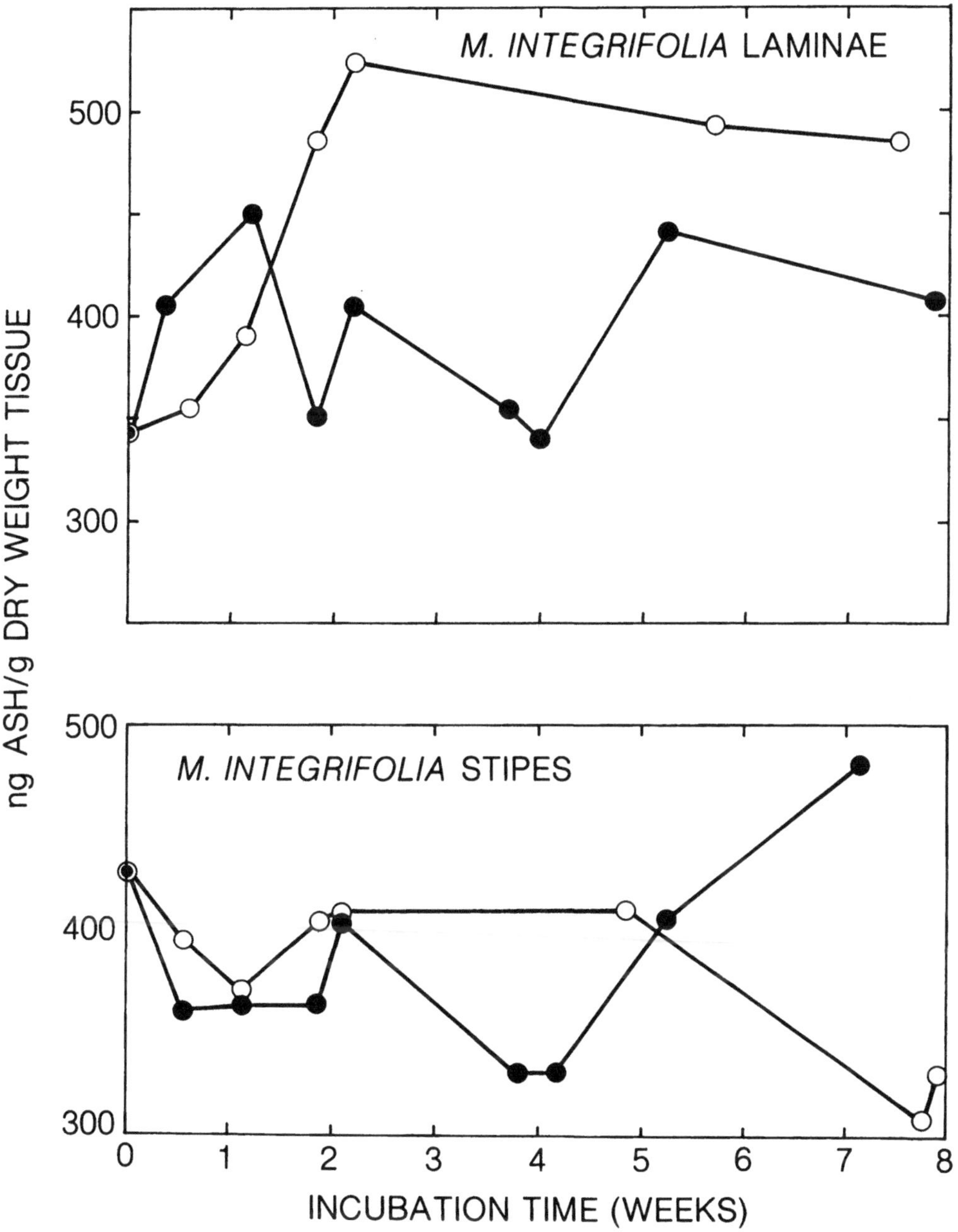

Fig. 8: Ash content of decaying M. integrifolia laminae and stipes incubated inter- (o) and subtidally (•).

decreased as degradation proceeded until the ratio for laminae detritus approached 10:1 and that for stipes was < 17:1. Thus, degradation of these tissues in litter bags, appears to increase their nutritive value. Other studies of decomposing macrophytes in the marine environment have shown similar patterns of relative nitrogen increases (11,17,24).

Acknowledgements

We acknowledge the financial support of Simon Fraser University for this project through its Programs of Excellence funding. Technical assistance of Caroline Grant is appreciated.

References

1. Bancroft, K., Paul, E.A., Wiebe, W.J.: Limnol. Oceanogr. 21, 473-480 (1976).
2. Bland, J.A., Brock, T.D.: Mar. Biol. 23, 283-292 (1973).
3. Chan, E.C.S., McManus, E.A.: Can. J. Microbiol. 15, 409-420 (1969).
4. Cundell, A.M., Mitchell, R.: Int. Biodeterior. Bull. 13, 67-73 (1977).
5. Druehl, L.D.: Phycologia 9, 237-247 (1970).
6. Druehl, L.D.: Can. J. Bot. 56, 69-79 (1978).
7. Fankboner, P.V., Druehl, L.D.: Experientia 32, 1391-1392 (1976).
8. Fenchel, T.M., Jorgensen, B.B.: In Advances in Microbial Ecology. Vol. I. (M. Alexander, ed.), Plenum, New York, pp. 1-58 1977
9. Floodgate, G.D.: In Marine Borers, Fungi and Fouling Organisms of Wood (E.B.G. Jones, S.K. Eltringham, eds.), Organization for Economic Cooperation and Development, Paris, pp. 117-123 1971
10. Fogg, G.E.: Oceanogr. Mar. Biol. Ann. Rev. 4, 195-212 (1966).
11. Heald, E.J.: The production of organic detritus in a south Florida estuary. Ph.D. dissertation, Univ. of Miami, Florida 1969
12. Hobbie, J.E., Daley, R.J., Jasper, S.: Appl. Environ. Microbiol. 33, 1225-1228 (1977).
13. Laycock, R.A.: Mar. Biol. 25, 223-231 (1974).
14. Leighton, D.L.: Nova Hedwigia 32, 420-453 (1971).
15. Morrison, S.J., King, J.D., Bobbie, R.J., Bechtold, R.E., White, D.C.: Mar. Biol. 41, 229-240 (1977).

16. Menzel, D.W., Vaccaro, R.F.: Limnol. Oceanogr. 9, 138-142 (1964).

17. Odum, E.P., de la Cruz, A.A.: In Estuaries, (G.H. Lauff, ed.), Amer. Assoc. Adv. Sci., Washington, D.C., pp. 383-388 1967

18. Rau, G.H.: Limnol. Oceanogr. 23, 356-358 (1978).

19. Robinson, J.D., Mann, K.H., Novitsky, J.A.: Amer. Soc. Limnol. Oceanogr., 2nd Winter meeting, Los Angeles, Calif., abstract (1980).

20. Russell-Hunter, W.D.: Aquatic Productivity: An Introduction to some Basic Aspects of Biological Oceanography and Limnology. Collier-Macmillan, London 1970

21. Scagel, R.F.: Pac. Sci. 15, 494-539 (1961).

22. Sibert, J., Brown, T.J., Healey, M.C., Kask, B.A., Naiman, R.J.: Science 177, 649-650 (1977).

23. Skerman, T.M.: N.Z. J. Sci. Technol. B. 38, 44-57 (1956).

24. Smith, B.D.: A qualitative and quantitative assessment of seaweed decomposition in the Strait of Georgia. M.Sc. thesis, University of British Columbia, Vancouver, B.C. 1979

25. Strickland, J.D.H., Parsons, T.R.: A Practical Handbook of Seawater Analysis. 2nd ed., Bull. Fish. Res. Board Canada 1972

26. Wood, E.J.F.: Marine Microbial Ecology. Reinhold, New York, 1965

DISCUSSION

WHEELER: Have you looked at degradation on the plant itself rather than in litter bags?

ALBRIGHT: No.

WHEELER: You can get a whole time series of decay using plants, which you may be able to compare with data from litter bags.

DRUEHL: A graduate student of mine, Bill Roland, looked at the colonization of everything from early bacteria to bryozoa on Macrocystis blades at different heights as a seasonal phenomenon and you get a classic pattern -- great differences between plants at different times of the year and with age.

SIBERT: Your 60% efficiency of conversion of kelp to bacterial biomass sounds pretty good to me. It agrees well with the laboratory data.

ALBRIGHT: It worries me a bit, 62.5% seems awfully high for efficiency of degradatious material.

SIBERT: It does, but that is the way bacteria are!

FOSTER: Could you explain the decrease in C/N ratio simply by a drop in carbon?

ALBRIGHT: The data show some drop in carbon and some increase in nitrogen, but not very much. The thing is if you get colonization by microorganisms I would expect them to bring a lot more nitrogen onto the plants colonized; probably non-protein nitrogen, such as in the cell wall of the bacterial cell, acetylglucoseamine for example. What I was wondering about is whether this nitrogen is available for nutrition, as some people working with seagrasses claim.

SCHMITZ: What do you understand by "crude fiber"?

ALBRIGHT: This is the assay of Strickland and Parsons (ref. 25) for crude fiber and it is pretty well cellulose.

SCHMITZ: Did you ever try to extract specifically for alginate?

ALBRIGHT: No.

SRIVASTAVA: You are using 1.5 mm hole bags, those are big holes for all kinds of things to be coming in and going out and you have no control over that. What is left inside may not be a true reflection of the trend from one week to the next. I have no solution for the problem, but this is a major factor of error. Maybe you should try to have bags of different pore sizes.

ALBRIGHT: We used litter bags because, it was convenient to do it that way, to get something going on the project. Eventually, we hope to use the Cepex-type bags, and not have to worry about losing detrital material. With Cepex-type bags we will try to retain everything to get an appreciation of kinetics of degredation.

CABOT: Did you look at any sectioned material to see whether bacteria were inside the cells?

ALBRIGHT: No. We looked at it with SEM. Bacteria would probably be one of the first things to invade the tissue as a primary colonizer. Probably, it would not be fungi. We did not notice particularly high levels of fungi, ever, in any of our degrading material.

CABOT: Would you say that they were all healthy blades; do you not have to cut them?

ALBRIGHT: They are healthy blades, but we have to cut them. This is another problem with this type of experiment. To get enough material of 100 grams into those bags, you have to cut small portions and you lose many things, excretions, photosynthate, initially that way.

McCONNAUGHEY: I would think that those holes in the bags would be large enough for harpacticoids to get in. Do you see any harpacticoids creeping around in your kelp?

ALBRIGHT: Towards the end of the experiment we did, particularly on the samples that were incubated in the intertidal region close to the sediment.

McCONNAUGHEY: There are harpacticoids which appear to be specifically adapted for invading seagrasses. You can see tham actually burrow into the seagrass blades.

CABOT: In some of our studies, we have seen gammarid amphipods sometimes associated with decaying Nereocystis.

Part IV. Polysaccharides, kelp farming and harvest

TOWARD IMPROVED UNDERSTANDING OF POLYSACCHARIDE SYNTHESIS AND STORAGE IN MARINE ALGAE

R.G.S. Bidwell

Genu Products Canada Ltd., and Atlantic Regional Laboratory, N.R.C., Halifax, N.S.
(Address: Wallace RR #1, Nova Scotia, Canada, BOK 1YO)

Introduction

The most important commercial compounds in marine algae at the present time are polysaccharides. The many individual applications of these carbohydrates depend on their very specific chemical properties. However, our present knowledge of the chemistry and physical chemistry of algal polysaccharides is inadequate to define the chemical parameters that underlie specific desirable or undesirable industrial properties. Furthermore, we have only limited knowledge of the biology of algal polysaccharide production (18,21): when they are made (both seasonally and diurnally), where in the plant, from what precursors, and how influenced (both qualitatively and quantitatively) by what chemical or environmental factors. Information on all these points is vitally necessary for developing the harvest strategy of a financially successful seaweed industry. The record of failures to date on both coasts of Canada and elsewhere attests to the danger of ignoring the need for basic research.

I wish to state first that I do not have the answers to questions that must be asked. However, in my experience with commercial seaweed cultivation during the past three years I have had an opportunity to see what is needed in the way of basic research for a successful industry, and how it should be tackled. Since some of the research is expensive and the rewards come only late in the game, it is necessary to establish priorities and a strategy for financing and actually doing the required research. My purpose in this paper is to provide an account of how I think this should best be done.

Synthetic and Degradative Processes in Marine Macrophytes

Good harvest or culture strategy requires methods that give the best sustained yield of the most desirable polysaccharides. Continuous cost-benefit analysis is necessary to determine the best possible strategy, rather than seeking the largest return from any given harvest tactic. This in turn requires knowledge about the range of biological and biochemical aspects of plant growth and development, including a number of questions about the polysaccharides. These fall into ten main categories, which are: (i) synthetic pathways, (ii) mechanisms of glycosylation and sulfation, (iii) storage sites, (iv) diurnal and annual timing of synthesis and storage, (v) effects of natural or controllable environmental factors, (vi) consequences of the physiological or developmental state of the plant, (vii) control by chemicals of the physiological and biochemical aspects of polysaccharide synthesis and storage, (viii) effects of harvest damage, (ix) post-harvest ripening techniques, and (x) selection of naturally or genetically synthesized high-yielding varieties.

Biological Research Program

The following section will provide a brief summary of the present status and future requirements of the ten questions mentioned above.

Synthetic pathways. At present quite a lot is known about the synthetic pathways and precursors of alginate and fucoidan, although the problem of the multiple inversions required to convert mannitol or glucose into fucose is still unresolved (Fig. 1). Research in several laboratories

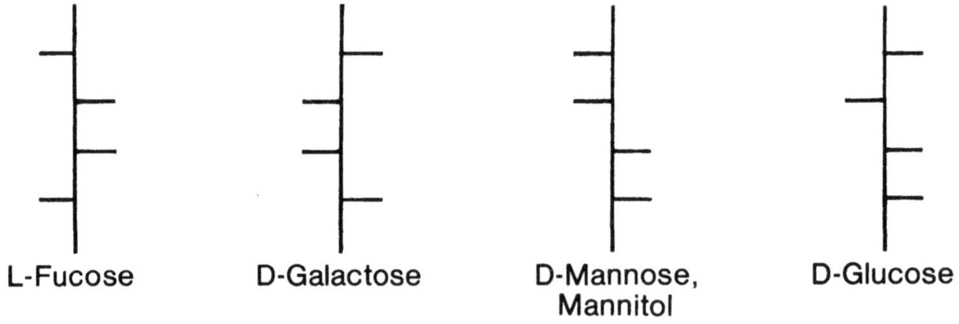

Fig. 1: Planes of symmetry of algal monosaccharides. Note that two, three or four inversions are necessary to convert D-sugars into L-fucose.

has shown that photosynthetic carbon is transported mainly as mannitol in brown algae, and that mannitol is the major substrate both of polysaccharide synthesis and of respiration (2,3,19,22). Finer biochemical focus, however, is lacking. Although this information is not of the highest priority, it will lead to a better understanding of the biochemical pathways to individual polysaccharides and the factors, both biological and environmental, that affect these pathways. The same starting material is used to produce different end products having different levels of commercial desirability. Understanding the switch or branch points may permit their control, as suggested in figure 2, with consequent improvement of the industrial value of the harvest.

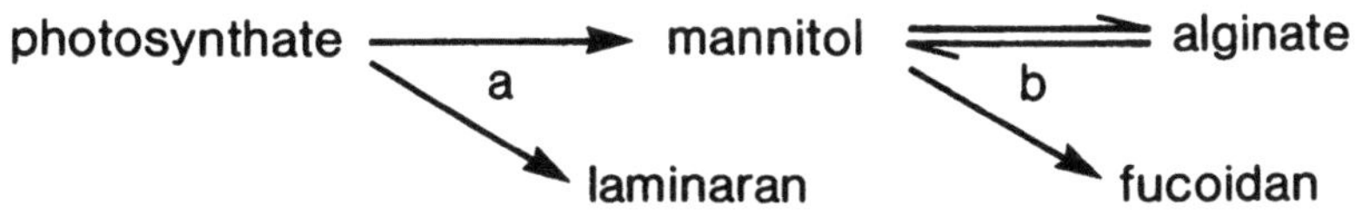

Fig. 2: Pathways of polysaccharide synthesis in brown algae (see refs. 6,10,14). Metabolic control at the switch or branch points (a,b) might lead to commercial possibilities for increasing the relative production of more valuable types of carbohydrate.

Mechanism of glycosylation and sulfation. While research in this area is not of the highest priority, some useful leads may come from better understanding of these processes. Recent work on transglycosylation suggests that the process in marine algae is not very different from that of higher plants (8). Work on the incorporation of sulfate suggests that this step follows polysaccharide formation, and so may be of greater interest (4,16). Because sulfation requires ATP, and hence light (which is the limiting factor for most algal photosynthesis), and because sulfation greatly affects the properties of polysaccharides, it may be of considerable interest to understand the process. As has been suggested by Hirst, sulfation may relate to dehydration or inversion in pre-formed polysaccharides such as fucoidan, carrageenan or agar (12). It will therefore be of interest to determine possible effects of short-term, perhaps intensive, treatment with sulfate or other chemicals designed to affect the sulfation of polysaccharides prior to harvesting.

Storage sites. Research in this category has much higher priority than in the previous ones because it relates to harvest strategy for continuous production. By and large, polysaccharides are highest in the young sections of the plant (13). For plants like *Laminaria* that grow from a basal meristem, this is not so good. However, it is most helpful for rockweeds and other plants that grow from the tip. It may be unimportant for a total-harvest technology, but it is of considerable concern where partial or sequential harvesting is appropriate.

Timing. Clearly, harvest timing depends primarily on plant growth and seasonal convenience. However, there are strong seasonal variations in the content and qualities of polysaccharides. For example, Stewart *et al.* (20) found that the carbohydrate content of *Ecklonia radiata* was higher in late summer than in spring by a factor of nearly two.

Seasonal effects also occur because different reproductive stages, which may be strongly seasonal, may make different types of polysaccharide (18). For example in the Gigartinaceae, as exemplified by *Chondrus crispus*, Chen *et al.* (7) showed that the tetrasporophyte makes lambda-carrageenan, whereas the gametophyte makes the commercially more desirable kappa-carrageenan. It would be of value to analyse both the qualitative and quantitative aspects of polysaccharides from brown algae during the seasonal growth of algal beds.

Environmental factors. The effects of natural or of controllable environmental factors have high priority for investigation. Diurnal variations may be of great interest. Photosynthesis adds considerable amounts of carbon to the plant each day. At what time of the day is this converted into polysaccharide? If this occurs immediately, as suggested by Hellebust and Haug for alginate (11), then harvesting is best done late in the day. If synthesis occurs overnight, as may be for some other polysaccharides (11,14), then continuous daytime harvesting is acceptable. On the longer term the effects of, for example, a week of sunless weather or an unnaturally warm spell are disastrous in terms of yield of cultured *Chondrus* (unpublished data). Comparable information on wild stands would be of value. Harvesting should clearly be tied to the weather. In addition, there is the possibility of affecting the quality and quantity of polysaccharides by fertilization. Surprisingly, chemical fertilizers added to the sea need not be wasted. Chinese commercial seaweed growers have experienced over 80% recovery of chemicals added to dense seaweed beds, which is better than most flatland agriculture (C.K. Tseng, personal communication).

Physiological or developmental state of plants. These effects may not be so great with browns, although they are known to be of great importance in red algae (7,18). However, little information is available about ways in which relative quantities of various polysaccharides in brown algae may change with development, or how the quantity or quality of any given polysaccharide is affected. Research on these points may be of considerable value in determining overall harvest strategy.

Chemical control. If there is inherent variation in the quality and quantity of polysaccharides, then it should be possible to control this to the harvester's advantage by the use of chemicals. As an example, the asparagine content of wheat leaves can be increased many-fold by appropriate brief treatment with nitrogen and the phytohormone, auxin. We used this chance observation as the basis of a commercial biosynthesis of labelled asparagine in the 1960's (17). Now, auxin and the other higher plant hormones may not work in brown algae. In fact, no one seems to know what are the chemical hormone systems of seaweeds. This field of plant physiology is wide open. Marine algae are so primitive that their developmental control systems may be quite different from those of higher plants. But control mechanisms they must have. When we discover them we shall be in a very powerful position to control algal growth and physiology. Certainly, the research needed will be basic, extensive and costly. Clearly, some way needs to be found to develop a research program to investigate these problems.

Harvest damage. This question is simple but important. Do the effects of partial harvesting increase or decrease polysaccharide production in the remaining portions of the plant? What are the effects of mechanical damage to the plants?

Post-harvesting ripening. These techniques have been applied to cultivated seaweeds in the past, but they could equally be applied to wild harvest. Some red algae such as Chondrus crispus can, by appropriate treatment, be made to increase their content of desirable carbohydrates by as much as 50% while suffering a net weight loss of only about 10% (unpublished data). Such treatments usually require that the plants be held under special conditions (usually not very demanding) in a simple container for a period of about one week. The technique could easily be applied to brown algae in holding tanks or ponds. Studies to determine the physiological and chemical treatment of brown algae for this purpose should be done.

Selection of productive varieties. An extensive program that has already yielded some very interesting results is now in progress for carrageenophytes at Genu Canada Ltd.'s Nova Scotia laboratory (15). The success of the program depends on new techniques for the rapid screening of carbon assimilation rates under conditions that permit accurate prediction of growth potential (5). These methods are based on the "artificial leaf" technique developed for chloroplasts and unicellular algae (1,9). Clones from promising plants are evaluated in growth chambers and outdoor cultivation facilities for productivity and carrageenan content. A selection program for brown algae is clearly of great importance for the future success of the harvest or cultivation industry. Since a number of species have now become susceptible of laboratory cultivation throughout their life cycle, the development of genetic programs for the improvement of growth rates and commercial desirability of these species is a logical next step with high priority.

Chemical research program. Beside biological research, much carbohydrate chemistry needs to be done. Two requirements are of special importance: (i) to relate chemical or physical-chemical structure to industrial application, and (ii) to develop rapid, simple techniques for evaluating large numbers of samples in a simple laboratory by technical personnel. Quantitative analytical procedures of this sort have been developed for carrageenans by Genu Products Canada Ltd., and work is now proceeding on the rapid analysis of industrial qualities. However, much solid basic chemical research is still needed in this field.

<u>Operation and funding.</u> In the previous sections I have recommended much basic and expensive research for an industry that can never be large or lucrative and which is known for its many past financial failures. Clearly a company devoted to harvesting kelps or cultivating agarophytes cannot realistically be expected to have the resources to set up programs for all the research I have outlined. The cost alone would be prohibitive, aside from the difficulties of adequate staffing. Yet the success of the venture may depend upon the outcome of these research programs. The solution to this dilemma depends upon the approach to research programs and the nature of their direction.

First, there are government and university laboratories fully set up and competent to do the required research, so no capital costs need be considered. However, it has been found unproductive merely to contract out research projects to such laboratories. The reason is that university and government scientists, by and large, do not appreciate the special problems and research needs of industry. Their priorities and concepts of timing are quite different. They are, in essence, ivory tower scientists, not industrialists.

The obvious alternative of hiring young scientists has been tried many times by small industrial concerns and found counterproductive. The newly graduated tend to be too concerned with their scientific reputation, and furthermore they lack the broad, comprehensive "synthetic" experience that comes from long years of practising research science. But hiring senior scientists on a full-time basis is obviously much too expensive.

The successful compromise is to generate think-tanks by forming loose, synthetic groups of senior scientists, together with engineers, market analysts and industrial leadership, on a consulting basis. The costs are relatively low, but the "brainstorm" effect of a group of experienced scientists in various fields together with design engineers and industrialists is very great. The output, per unit cost, exceeds by one to several orders of magnitude the usual output of either a conventional government or university laboratory or an industrial research unit. Actual research is done by technicians housed in appropriate government or university laboratories, and costs can usually be largely covered by grants from various government sources.

The important distinction in this system is that the overall direction of the research program comes from industry. Final decisions in all phases of the scientific program are made by the engineering-industrial leadership of the think-tank, not by the scientists. This is essential, because the industrialists are ultimately responsible for the success or failure of the project, and because scientists normally do not easily grasp the industrial and engineering requirements. On the other hand, it is the scientists that establish the validity and value of specific research priorities for the industrial leadership. This way, the most useful and important research gets done first, and research lines, however interesting, that are not leading to advantageous cost benefits can quickly be dropped or shelved.

References

1. Bidwell, R.G.S.: Can J. Bot. 55, 809-818 (1977).
2. Bidwell, R.G.S., Craigie, J.S., Krotkov, G.: Can. J. Bot. 36, 581-590 (1958).
3. Bidwell, R.G.S., Ghosh, N.R.: Can. J. Bot. 40, 803-807 (1962).
4. Bidwell, R.G.S., Ghosh, N.R.: Can. J. Bot. 41, 209-220 (1963).
5. Bidwell, R.G.S., McLachlan, J.L., Lloyd, N.D.H.: Proc. Can. Soc. Plant Physiol. 20 (in press).
6. Bidwell, R.G.S., Percival, E., Smestad, B.: Can. J. Bot. 50, 191-197 (1972).
7. Chen, L.C-M., McLachlan, J., Neish, A.C., Shacklock, P.F.: J. mar. biol. Ass. UK 53, 11-16 (1973).
8. Coughlan, S., Evans, L.V.: J. Exp. Bot. 29, 55-68 (1978).
9. Dodd, W.A., Bidwell, R.G.S.: Nature 234, 45-47 (1971).
10. Hellebust, J.A., Haug, A.: Can. J. Bot. 50, 169-176 (1972).
11. Hellebust, J.A., Haug, A.: Can. J. Bot. 50, 177-184 (1972).
12. Hirst, E.L.: Proc. Chem. Soc., July 1958, pp. 177-187 (1958).
13. Kremer, B.P.: Helgoländer wiss. Meeresunters. 27, 115-127 (1975).
14. Leppard, G.G.: Can. J. Bot. 52, 773-781 (1974).
15. Lloyd, N.D.H., McLachlan, J.L., Bidwell, R.G.S.: Proc. Int. Seaweed

Symp. 10, (in press).

16. Loewus, F., Wagner, G., Schiff, J.A., Weistrop, J.: Plant Physiol. 48, 373-375 (1971).

17. Maxwell, M.A.B., Bidwell, R.G.S.: Can. J. Bot. 48, 923-928 (1970).

18. McCandless, E.L., Craigie, J.S.: Ann. Rev. Plant Physiol. 30, 41-53 (1979).

19. Nicholson, N.L., Briggs, W.R.: Am. J. Bot. 59, 97-106 (1972).

20. Stewart, C.M., Higgins, H.G., Austin, S.: Nature 192, 1208 (1961).

21. Turvey, J.R.: In Biochemistry of Carbohydrates, Vol. II, (D.J. Manners, ed.) pp. 151-177. University Park Press, Baltimore, MD 1978

22. Yamagouchi, T., Ikawa, T., Nisizawa, K.: Plant Cell Physiol. 7, 217-229 (1966).

DISCUSSION

DRUEHL: There is one aspect that you did not mention, that is, industry helping industry. I am now alluding to the fact of positive exchange between various companies, and how the industry is going to grow, as a group? Or does the industry want everybody to reproduce everything over and over?

BIDWELL: First, I must mention that the post-harvesting ripening system was patented by a government laboratory, by the NRC, not by any industrial concern. In fact, there are two companies in Nova Scotia now that are using the cultivation and various other aspects of the system that Neish established and both of these have bought rights on the patent. Now as far as the industry helping industry, there are some reasons why they do not tend to be quite so friendly about cooperating as you might think they should be. Obviously, if you can get there six months ahead of the other fellow you are quite a bit ahead in terms of your development. Also, if you spend your investor's money on doing some research, you really do not want to give it to others who have not spent that money. That is the kind of attitude you face in industry and it is one that some of us at Genu are trying to break down. It has now got to the stage, as you say, that anybody's failure is everybody's failure, and anybody's success is everybody's success. So I think it is important that the industrial concerns should get together as far as possible.

WILLENBRINK: You mentioned in the first part of your lecture that you could hopefully change the direction of the overall biosynthesis. Did you mean by changing the genes, by using recent techniques in genetics and other ways?

BIDWELL: First, I must stress the word 'hopefully.' Second, I am not quite sure why, but there are ways, many situations that relate to environmental factors which will cause a massive shift in metabolism. An example is the asparagine synthesis system in wheat leaves where sucrose is converted into asparagine. There may be other kinds of events, that seem to be mediated by environmental factors, probably through hormone-type mechanisms, which in turn act to control these enzyme systems. There are any number of possible mechanisms: either by enzyme induction or by enzyme control, or even by the short distance translocation or transport of substrates to make them accessible, say from one compartment to another, to make them accessible to different synthetic systems. I was really thinking of that sort of control. I do not know how it would work or what could be done, but some experiments might give us some leads.

SRIVASTAVA: You have the arrows going both ways, mannitol to alginate (Fig. 2). Do you have any basis for that?

BIDWELL: Some time ago, we did some pulse chase experiments with ^{14}C and found that the alginate fraction was actually turning over, whereas the fucoidan and laminaran, at least in the relatively short term experiments, that is hours or days, that we did, seemed to be end products.

WILLENBRINK: I am going to add another question. Is it not true.that only the last step for the mannitol biosynthesis is completely restricted to mannitol and the step before leads also to alginate? I think the alginate synthesis is not via mannitol.

BIDWELL: I am not quite sure. But, we did find that mannitol and alginate were being interconverted rather readily.

FARMING *MACROCYSTIS* AT COASTAL AND OCEANIC SITES

W. North, V. Gerard, and J. Kuwabara
California Institute of Technology, Kerckhoff Marine Laboratory
101 Dahlia Street, Corona del Mar, California, USA

Since California kelp beds first attracted attention as a source of potash in 1910, commercial interest in the giant kelp, *Macrocystis*, has expanded considerably. Today, approximately 150,000 wet tons of kelp, primarily *Macrocystis*, are harvested annually in California and utilized for production of alginates and other lesser products. Actual crop size varies from year to year, but even the largest crops do not provide sufficient raw material to meet current demands of industry. Furthermore, interest has recently arisen in possible usage of *Macrocystis* biomass as a feedstock for methane production. There are thus needs not only for enhancing productivity of existing *Macrocystis* populations, but also for expanding the resource by development of oceanic farms. Evidence suggests that both of these objectives hinge on maintenance of optimal nutrient conditions for support of maximum growth.

Kelp nutritional requirements *vs.* nutrient availability

In order to meet this objective, either in coastal kelp forests or oceanic farms, nutritional requirements of the plant must first be defined. The elemental requirements for development and growth of *Macrocystis pyrifera* from settled spores through the gametophytic stage to embryonic sporophytes have recently been determined (1). Seven micronutrients and two macronutrients were necessary additions to the artificial seawater medium (Aquil) for successful development (Table I). Little is known of micronutrient availability in coastal waters. However, addition of the two macronutrients, nitrate and phosphate, to surface water collected from nearshore sites usually stimulated growth of juvenile *Macrocystis* sporophytes in the laboratory (North, unpublished data). Plots of

Synthetic and Degradative Processes in Marine Macrophytes

Table I: Nannomoles of nine nutrients and of EDTA added to the completely defined artificial sea water Aquil to yield a medium that sustained development by Macrocystis zoospores completely through the gametophyte stage to multicellular embryonic sporophytes in 12 culturing days.

Nutrient used	nM added	Free ion conc. (nM)	Major species present and (%)	
Fe^{+3}	3.5×10^2	7×10^{-11}	FeEDTA	(100)
Mn^{+2}	30	3	MnEDTA	(65)
			$MnCl^+$	(23)
Co^{+2}	70	7×10^{-2}	CoEDTA	(99)
Cu^{+2}	10	4×10^{-5}	CuEDTA	(99)
Zn^{+2}	1.7×10^2	7×10^{-2}	ZnEDTA	(100)
MoO_4^{-2}	1.0×10^2	1×10^2	--	
$EDTA^{-2}$	6.0×10^3	7×10^{-5}	CaEDTA	(89)
			FeEDTA	(6)
NO_3^{-1}	2.0×10^4	2×10^4	--	
PO_4^{-3}	2.0×10^3	3×10^{-1}	HPO_4^{-2}	(51)
			$MgHPO_4^{-1}$	(47)
I^{-1}	1.0×10^2	1×10^2	--	

phosphate vs. nitrate (in this paper "nitrate" is used to denote nitrate (NO_3) and nitrite (NO_2), ed.) concentrations from inshore sea water samples show that small amounts of phosphate remain even when nitrate falls to near undetectable levels (Fig. 1), indicating that nitrogen becomes limiting to growth before phosphorus.

Macronutrient concentrations in the coastal waters of Southern California fluctuate in a seasonal pattern. Natural upwelling of nutrient-rich water from depths of 100-300 m or more occurs typically during the spring and early summer. During the upwelling season, nitrate levels are high throughout most of the offshore water column (Fig. 2). Significant upwelling is rare during the late summer, fall, and winter, and nitrate concentrations are typically low, especially in offshore surface water. Temporal fluctuations in dissolved phosphate closely follow those of

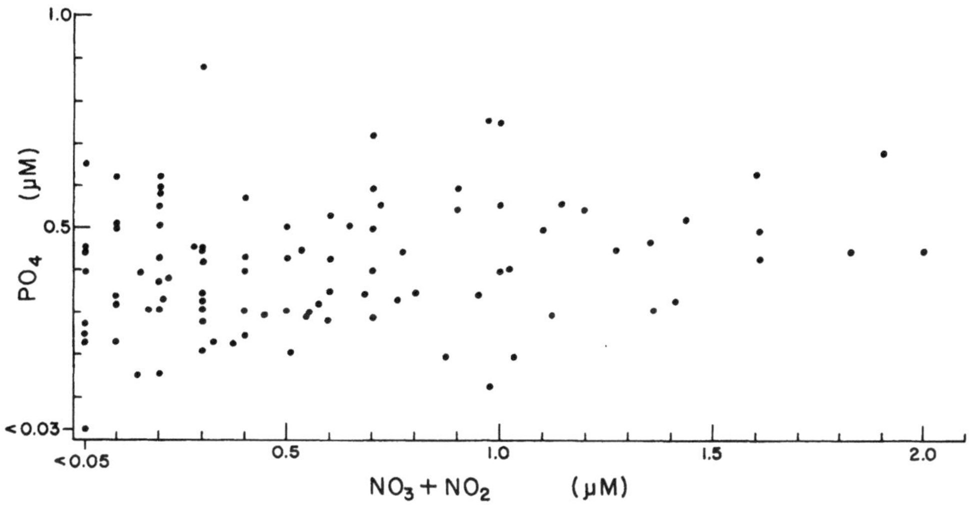

Fig. 1: Relationship between dissolved nitrate and phosphate in nutrient-poor inshore water. Concentrations of NO_3+NO_2 and PO_4 are shown for seawater samples collected off Corona del Mar, California. Data are shown only for samples with $NO_3+NO_2 \leq 2$ µM.

nitrate, but concentrations are about one order of magnitude lower for phosphate. Inshore nutrient conditions are also affected by upwelling. High nitrate concentrations, up to 18 µM, occurred off Corona del Mar during spring and early summer 1979 (Fig. 3). An unusual upwelling occurred in fall 1979. Although this was not an intense event and lasted only a short time, nitrate concentrations were unseasonably high in both offshore and inshore waters (Figs. 2 and 3). Another major source of coastal nutrients is terrestrial runoff, especially during winters of heavy rainfall. During the particularly wet winter of 1980, nitrate concentration reached more than 30 µM off Corona del Mar (Fig. 3). The previous winter had only average rainfall, and nitrate did not exceed 4 µM. Vertical stratification was evident even at this shallow (10 m) site: bottom water became richer in nitrate than surface water during upwelling, but the opposite was true during the runoff period. Overall, 1979-80 was a year of high macronutrient supply, and inshore nitrate concentrations did not remain below 1 µM for more than one month of any season.

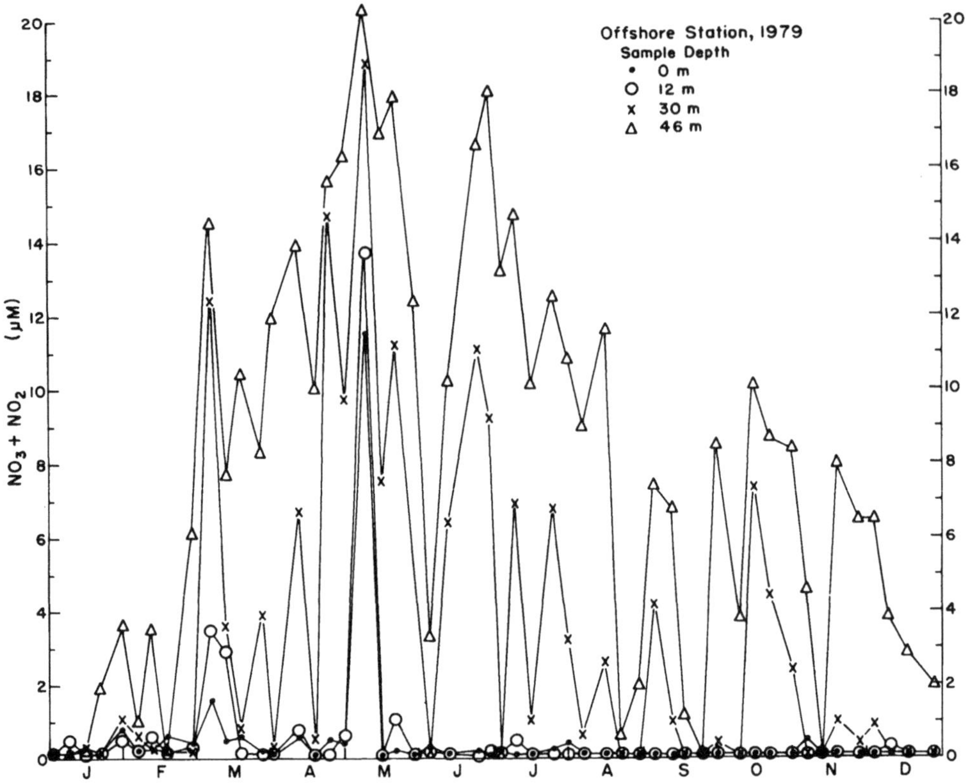

Fig. 2: Dissolved nitrate and nitrite concentrations in sea water samples collected at 0, 12, 30, and 46 m depths during 1979. The sampling site was located at the 550 m depth contour about 5 km south of Corona del Mar, California.

Kelp response to varying nutrient conditions

The influence of varying nutrient conditions on changes in total nitrogen content of lamina tissues of juvenile M. pyrifera sporophytes in laboratory experiments is shown in figure 4. In the field, during the upwelling season in 1979, nitrogen in blades near the bottom of the canopy fronds of adult plants at Corona del Mar rose from 1 to 3% (Fig. 3). Subcanopy and canopy blades also had relatively high nitrogen contents during upwelling, but the increase was smaller than among bottom blades which experienced greater exposure to nitrogen-rich water. Nitrogen

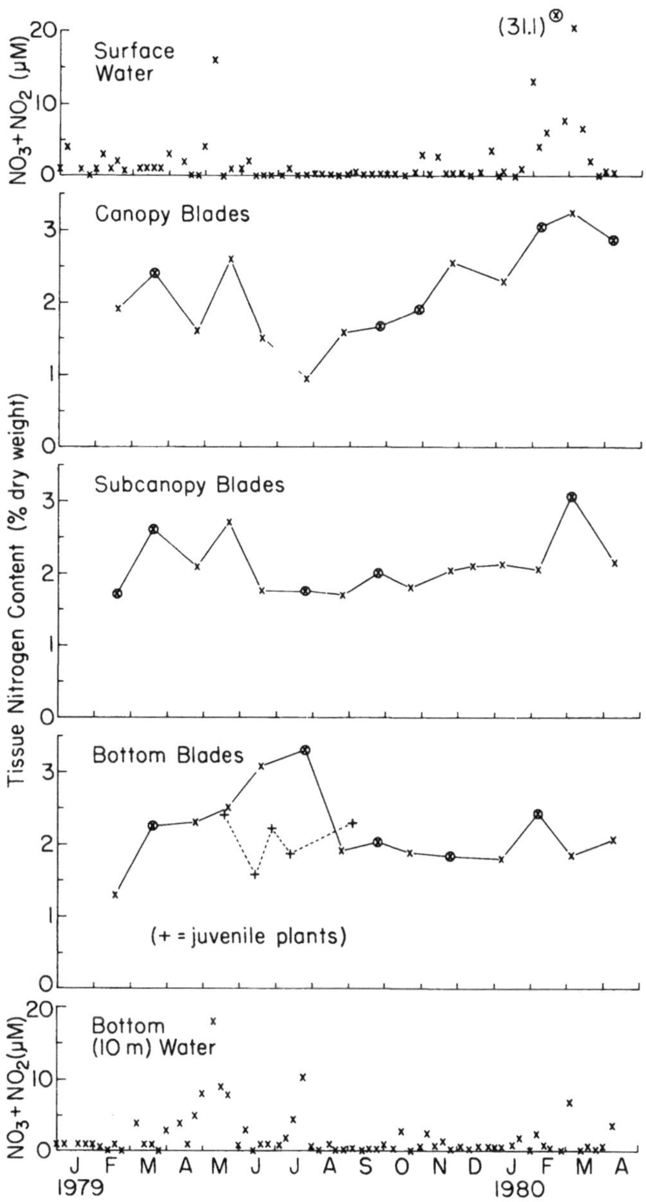

Fig. 3: Dissolved nitrate and nitrite concentrations in surface and bottom (10 m) sea water samples and nitrogen (total N) content of kelp canopy, subcanopy, and bottom blade tissue collected inshore at Corona del Mar, California. Each tissue nitrogen value is a mean of six individual blade samples collected from three fronds. Circled points indicate range > 1.

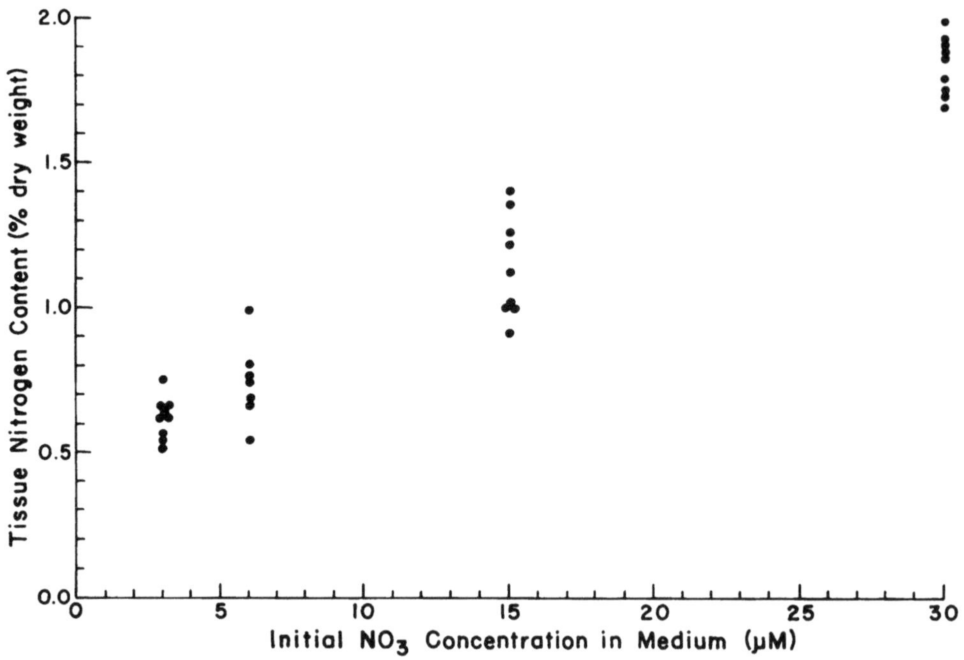

Fig. 4: Tissue nitrogen (total N) contents of juvenile Macrocystis sporophytes grown for 35 days under 4 different nutrient conditions in the laboratory. The culture medium consisted of 40 liters of offshore surface sea water with nitrate added to give initial concentrations between 3 and 30 µM. Phosphate was also added at an N/P ratio of 15/1. Plants were transferred to fresh medium every 2 days. Points at any one concentration represent individual plants.

content of juvenile sporophytes at the same site did not reach as high a level during upwelling as the bottom blades of adult plants, despite similar environmental conditions (Fig. 3).

Blades near the apex of canopy fronds had the most variable nitrogen content. During periods of heavy runoff and plentiful surface nutrients, these canopy blades showed the highest nitrogen levels (Fig. 3). Their nitrogen content dropped below 1% only once in 16 months, surprisingly at the end of the upwelling season. Experimental results suggest that tissue nitrogen below 1-1.5% of dry weight indicates nitrogen limitation of growth of both juvenile and adult Macrocystis sporophytes in the absence of external supply (North, unpublished data). Low nitrogen content in apical portions of an adult frond does not necessarily indicate nitrogen-limited growth conditions, however. Nitrogen contents in excess of 1.5% of dry weight in deeper frond portions may represent stored reserves which could be translocated within the frond as amino acids in sieve tube sap (2). Therefore, in the absence of external nitrogen supply, apical frond growth may be supported by translocation of internal reserves. During 1979-80, average nitrogen content of subcanopy and bottom blades at Corona del Mar generally remained around 2%. Nitrogen levels never approached 1% simultaneously in more than one of the three blade types collected during each sampling. Overall, there was no indication of nitrogen acting as a growth-limiting nutrient. However, 1979-80 was a period of unusually intense upwelling and heavy rainfall in Southern California. In years of average or inappreciable upwelling and runoff, low dissolved nitrogen supply probably does become limiting to kelp growth for periods of variable duration.

Suboptimal nutrient conditions may affect productivity of coastal kelp forests not only by limiting frond growth, but also by influencing frond initiation rates and frond longevity. Extreme nitrogen deprivation appears to cause weakening and sloughing of frond tissues. <u>Macrocystis</u> canopies declined severely along the Southern California coast during exceptionally nutrient-poor periods, such as fall and winter 1976-77. Less extreme deprivation may influence frond longevity to a lesser but measurable extent. Any factor which influences frond initiation and longevity must ultimately determine frond density, standing crop, and harvest yield. The problems then are to identify oceanographic conditions under which nitrogen becomes limiting to kelp productivity and, at the same time, determine what parameters (e.g. frond growth rate, longevity, etc.) are significantly affected. Finally, we need to know in what ways, if any, the limitation may be overcome efficiently.

Fertilizing coastal kelp beds

We approached these problems by conducting a large-scale fertilizing experiment. During four intervals in 1979-80, an experimental kelp forest at Leucadia, San Diego County, was fertilized with a mixture of ammonium sulfate and ammonium phosphate dispersed as crystals from a crop-dusting

Table II: Summary of fertilizing and harvesting data from experimental kelp bed at North Leucadia, 1978-1980.

First harvesting Nov 29-30 1978
375 wet tons obtained
computed area harvested - 0.08 km^2

First fertilizing Jan 8 to 24 1979
13 tons ammonium sulfate dispersed

Second harvesting Feb 28 1979
203 wet tons obtained
Computed area harvested - 0.106 km^2

Second fertilizing April 9 to 24 1979
12 tons ammonium sulfate + 1 ton ammonium phosphate dispersed

Third harvesting May 23, May 31, June 1 1979
399 wet tons obtained
Computed area harvested - 0.009 km^2
This is probably a poor estimate of area harvested because the preharvest photo was taken on May 8, 15 days before the harvesting began.

Third fertilizing Sept 17 to Oct 3 1979
10 tons ammonium sulfate dispersed + 0.4 tons ammonium phosphate

Fourth harvesting Nov 29 to Dec 10 1979
1104 wet tons obtained
computed area harvested - 0.102 km^2

Fourth fertilizing Jan 25 to Feb 28 1980
15 tons ammonium sulfate + 0.6 tons ammonium phosphate dispersed

helicopter (Table II). A variety of preliminary data were collected in order to maximize effectiveness of this treatment. Firstly, we chose ammonium over nitrate as a nitrogen source, because uptake of the former occurs more rapidly, especially at concentrations greater than 20 µM (3). Brief exposures (1-10 min) of juvenile Macrocystis sporophytes to concentrations up to 1 M ammonium sulfate resulted in temporary adverse effects on growth rate and tissue turgor, but not in permanent damage to the plants. Calculations indicated that nitrogen uptake efficiency increases in direct proportion to biomass concentration. A model of vertical distribution of biomass for a Macrocystis population showed that 50% or more of the kelp is typically found as canopy in the upper 1 m of the water column (P.D. Kirkwood, unpublished data). Therefore, we

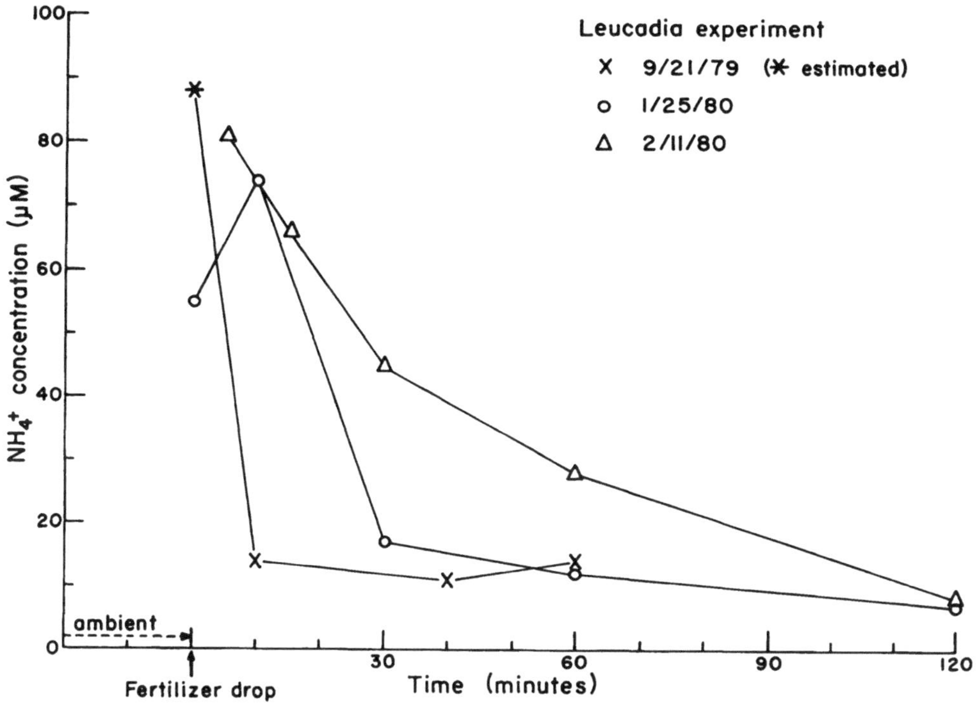

Fig. 5: Concentrations of dissolved ammonium in interstitial sea water samples collected from the kelp canopy at Leucadia before and after fertilizing.

selected crop dusting as a dispersal method which would concentrate fertilizer in the kelp canopy region. Ammonium sulfate crystals weighing 0.7 mg or less dissolved completely in stagnant sea water in less than 30 sec while falling through vertical distances of less than 1 m, so 0.7 mg was determined as the maximum desired crystal size. Finally, a few simple dye dispersal studies showed that dense kelp canopies act as barriers to currents. Water flow within the canopy area was much reduced in comparison to flow in open water, and we assumed that increased residence time of fertilizer dispersed into interstitial canopy water would lead to relatively high uptake efficiency.

During each fertilizing interval, one ton of fertilizer was dropped on the 0.15 km^2 canopy area of the Leucadia kelp bed daily for two weeks or every other day for one month. Uniform dispersal should have resulted in

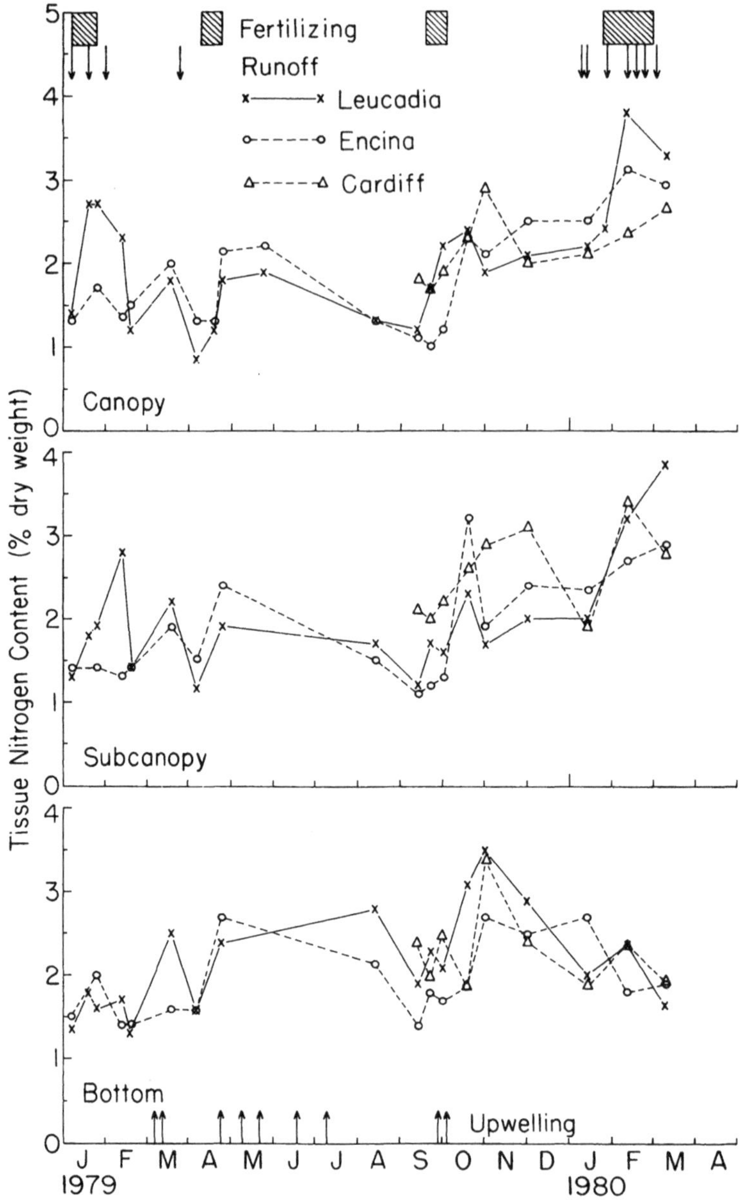

Fig. 6: Nitrogen content (total N) of canopy, subcanopy, and bottom blade tissues collected from experimental kelp plants at Leucadia and from control plants at Encina and Cardiff, California. Each tissue nitrogen value is a mean of six individual blade samples collected from three fronds. Arrows at bottom and top of graphs indicate periods of significant upwelling and runoff. Bars at top indicate intervals of fertilizing at Leucadia.

maximum ammonium concentrations approximately 50 µM above ambient levels which ranged between 0.7 and 2.3 µM. Interstitial water samples collected within the kelp canopy before and after fertilizing showed that actual ammonium concentrations were initially elevated 50-90 µM over ambient and remained higher than ambient for at least two hours (Fig. 5). Nitrogen content of the canopy blades at Leucadia characteristically rose during the fertilizing intervals (Fig. 6). Unfortunately, three of the four intervals coincided with significant upwelling or runoff, so tissue nitrogen levels increased simultaneously at the two control sites located 6 km to the north (Encina) and south (Cardiff). Compared to the controls, experimental plants clearly showed higher nitrogen levels in canopy and subcanopy blades in only one instance, during fertilizing in January 1979. Comparisons of frond growth and initiation rates at the three sites

Table III: Apical growth rates of canopy fronds (*Macrocystis pyrifera*) at three kelp beds near Leucadia, California. The experimental bed at Leucadia was fertilized with ammonium sulfate from September 17 through October 3, 1979. The control beds at Encina and Cardiff were not fertilized. The two measurement periods represented five days prior to fertilizing and five days at the end of the fertilizing interval at Leucadia. T-tests were used to determine significance levels of differences between growth rates during the two measurement periods.

	Measurement Period					
	Sept 7 - Sept 12			Sept 27 - Oct 2		
Location	Ambient Nitrate (µM)	No. of fronds	Mean Growth Rate cm/day	Ambient Nitrate (µM)	No. of fronds	Mean Growth Rate cm/day
	Fronds with apical blades at final measuring					
Leucadia	0.3	12	5.9 (±1.6)[a]	5.7-6.8	5	10.0 (±2.0)[a]
Encina	0.1	14	4.0 (±1.7)[b]	6.0-8.5	12	9.2 (±1.3)[b]
Cardiff	1.4	20	5.8 (±1.5)[b]	4.7-5.2	12	8.9 (±1.4)[b]
	Fronds with terminal blades at final measuring					
Leucadia	-	11	3.3 (±1.3)[b]	-	21	6.9 (±1.0)[b]
Encina	-	9	1.9 (±0.9)[b]	-	15	6.0 (±1.4)[b]
Cardiff	-	7	2.8 (±0.9)[a]	-	15	4.9 (±1.2)[a]

[a] Significance level - $p < 0.05$.
[b] Significance level - $p < 0.01$.

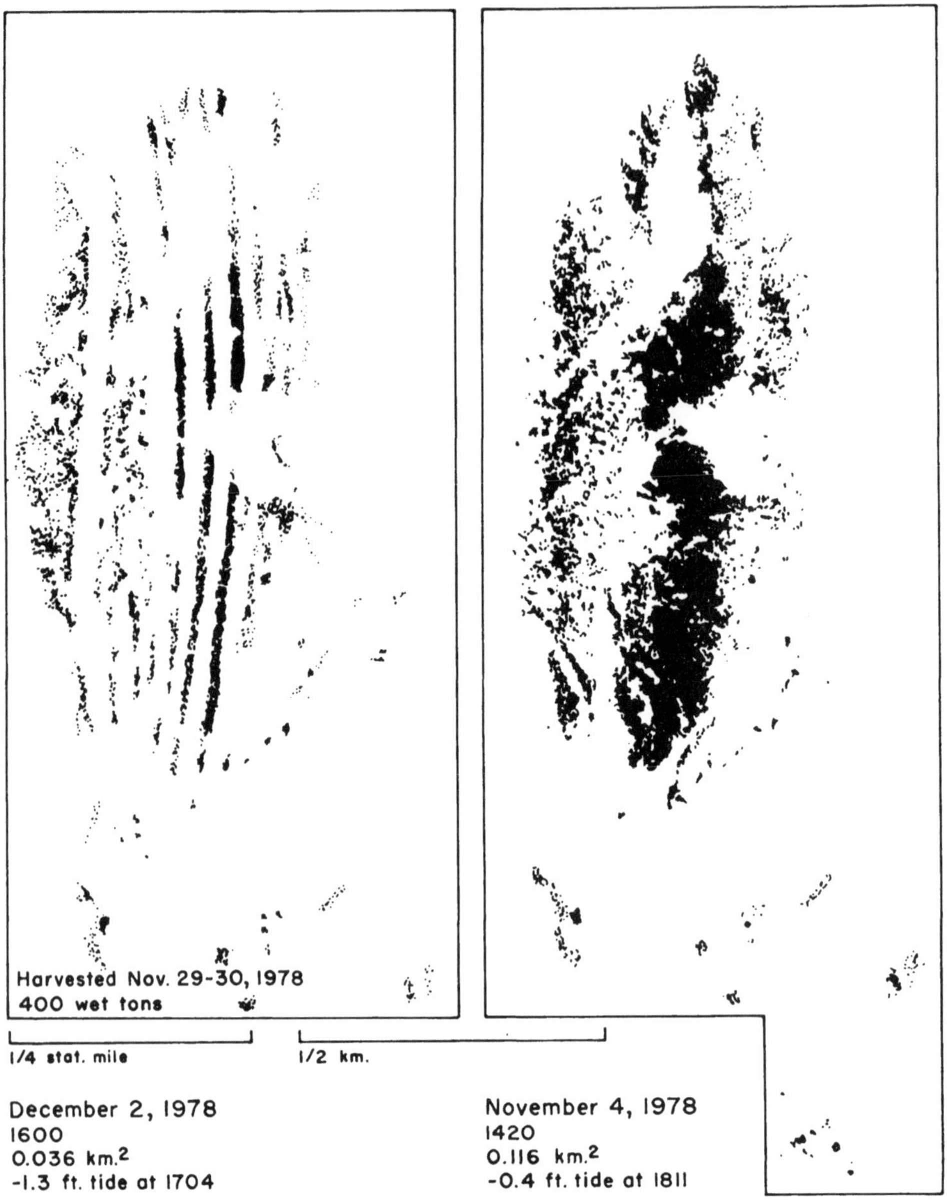

Fig. 7: Examples of drawings used to compute area of kelp canopy removed by harvesting at Leucadia and control sites. Drawings were made from aerial photos taken before and after harvesting.

during fertilizing showed no significant differences, presumably because of the high ambient nutrient conditions. However, one set of growth data did implicate nitrogen as a growth-limiting factor: growth rates increased significantly at all three study sites between mid and late September 1979 (Table III). The low growth rates occurred in mid September when ambient nitrate levels were low at the three sites and nitrogen contents in canopy and subcanopy blades at two of the sites averaged only 1%. High growth rates occurred in late September, when a strong unseasonal upwelling event coincided with the third fertilizing interval and raised ambient nitrate levels at all three sites.

The experimental bed at Leucadia was harvested by Kelco Company four times between November 1978 and December 1979. Yields were computed by dividing tons of kelp harvested by the area of canopy removed by the harvester. Area of canopy removed was taken as the area of the bed before harvesting less the area after harvesting. These areas were measured using drawings made from aerial photos of the bed (Fig. 7). Total yield from the entire period of fertilizing was over 2000 wet tons (Table II) or about 10-12 dry tons/acre/year (1DTAY = 2.2 metric tons/hectare/year). This yield compared to an average of approximately 0.4 DTAY for the annual California kelp harvest and to 1-4 DTAY for a highly productive bed at Pt. Loma (C. Barilotti, personal communication). However, a single harvest of the Cardiff control bed in December 1979 also produced a high yield. The results of our fertilizing experiment have been thus far inconclusive with respect to effect on yields, presumably because 1979-80 was a period of unusually high ambient macronutrient levels. Although nutrient enrichment of coastal kelp forests may not be an efficient means of enhancing productivity during high nutrient periods, evidence to date suggests overall that fertilizing will be effective in less opulent times.

Kelp nutrition at oceanic sites

The possibility of growing Macrocystis as an energy crop on oceanic farms depends critically on the development of cost- and energy-efficient fertilizing techniques. Oceanic surface water is low in macronutrients

(Fig. 2), and early attempts to grow kelp on offshore structures resulted in low growth rates and low tissue nitrogen levels, except during periods of natural upwelling (4,5). Offshore surface water enriched with nitrate and phosphate supported good growth of juvenile Macrocystis sporophytes in the laboratory (5), but high monetary and energy costs prohibit the use on oceanic farms of the fertilizing techniques we are utilizing for coastal forests. Artificial upwelling of nutrient-rich deep water provides a possible alternative. Below depths of about 300 m, macronutrient levels are consistently high (ca. 25-35 $\mu M\ NO_3$ and 1-3 $\mu M\ PO_4$ in the Southern California Bight). Juvenile sporophytes held in deep water in the laboratory grew well, but even better growth was obtained with a 50:50 mixture of deep and surface water (4). Results of micronutrient assays and enrichment experiments indicated that certain trace elements, Mn^{+2} and possibly Fe^{+3} for example, sometimes became limiting to kelp growth in

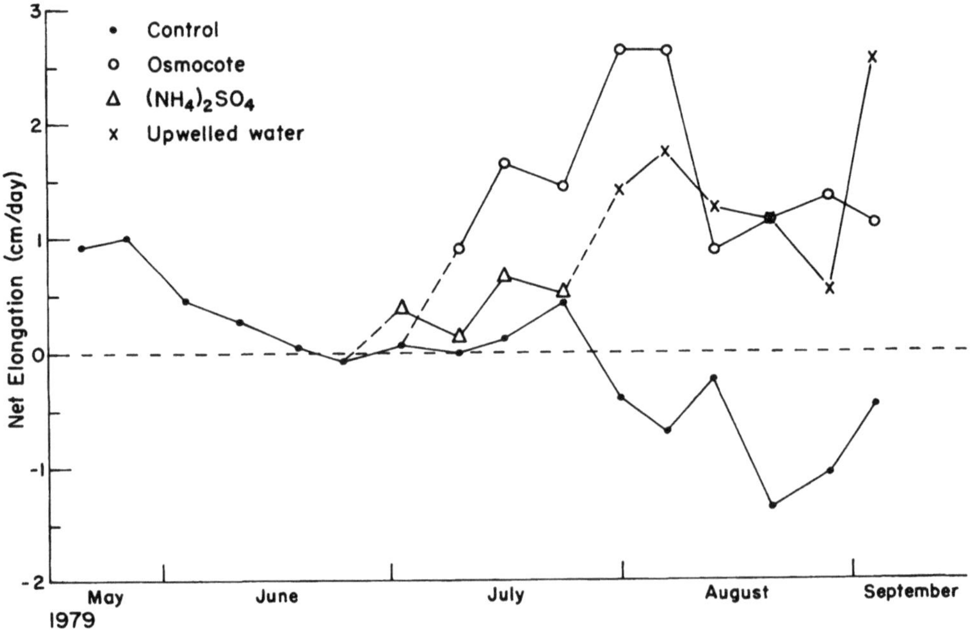

Fig. 8: Net elongation rates of juvenile Macrocystis sporophytes located on the offshore farm structure. Rates are shown for unfertilized control plants and for plants fertilized by three different methods. Each value is a mean of growth measurements for 4-28 individual plants.

pure deep water (5). The micronutrient compositions of deep and surface water appeared to complement each other, resulting in high growth rates.

One important objective of the Marine Farm Project is to measure yields from adult Macrocystis plants continually fertilized with artificially upwelled deep water in an offshore environment. For this purpose, a specially designed structure was constructed and installed in September 1978 at the 550 m depth contour about 7 km offshore from Laguna Beach, California. During November and December 1978, 103 adult kelp plants were transplanted to the farm and supplied with 10,000 gal/min of water pumped from a depth of 450 m. The transplants were destroyed within two months, due to strong currents causing entanglement and abrasion with the farm structure. However, more than 40,000 juvenile Macrocystis sporophytes were recruited onto the farm the following spring, presumably from spores liberated by the original adult transplants. We measured net elongation rates of the juvenile plants, some of which were fertilized by several techniques and others which remained unfertilized as controls. While growth of the control plants declined in July after the end of the upwelling season, plants exposed to artificially upwelled deep water maintained high growth rates (Fig. 8). From August through November, tissue nitrogen content of the fertilized plants averaged more than 2% of dry weight; tissues of the control plants averaged only 0.6-1.5% nitrogen. Other plants situated adjacent to a slow-release fertilizer (Osmocote) grew similarly to plants fertilized with deep water. Plants exposed to high concentrations of ammonium sulfate for brief intervals did not grow significantly faster than the control group (Fig. 8). Presumably, exposures were too brief and infrequent to allow adequate nitrogen uptake for maintenance of high growth rates. These results support the hypothesis that low macronutrient availability is the primary factor limiting kelp growth in offshore environments, and suggest that artificial upwelling will eventually prove to be a viable solution to the problem of maintaining optimal nutrient conditions on oceanic farms.

Acknowledgements

Current research support from the Office of Sea Grants under Grant No. 04-5-158-13, from Kelco Company, from Union Oil Company, and from the U.S. Department of Energy under contract E(04-3)-1275, is gratefully acknowledged, as is past support from the U.S. Navy and the National Science Foundation. We thank the staff of the Kerckhoff Marine Laboratory for assistance in all aspects of the work. Cooperation from Kelco Company in harvesting the experimental bed at Leucadia was crucial for acquiring yield data. In part, this work is the result of research sponsored by NOAA Office of Sea Grants, Department of Commerce. The U.S. government is authorized to produce and distribute reprints for governmental purposes notwithstanding any copyright notation that may appear.

References

1. Kuwabara, J.S.: Micronutrient requirements for Macrocystis pyrifera (L.) C.A. Agardh (giant kelp) gametophytes determined by means of a chemically defined medium, Aquil. Ph.D. thesis, California Institute of Technology, Pasadena, CA. 1980

2. Schmitz, K., Srivastava, L.M.: Plant Physiol. 63, 995-1002 (1979).

3. Haines, K.C., Wheeler, P.A.: J. Phycol. 14, 319-324 (1978).

4. North, W.J.: In Biological Solar Energy Conversion. (A. Mitsui, S. Miyachi, A. San Pietro, S. Tamara, eds.) pp. 347-361, Academic Press, New York 1977

5. North, W.J.: Proc. Ann. Symp. Fuels from Biomass 2, 999-1040 (1978).

DISCUSSION

DRUEHL: One aspect you did not mention in your explanation of optimal tissue for uptake was time. We have some evidence that suggests that for *Laminaria* ammonium uptake is doubled at night over day. Have you looked at the time factor?

NORTH: No, We have not looked at uptake during the day *vs.* night. I do not think my crew would get out there at night. Certainly, in an ocean farm where you have the upwelling going on continuously that could be a big factor and it should be studied.

SRIVASTAVA: Is there any way of fertilizing ocean farms while at the same time controlling the sea urchin population?

NORTH: The sea urchin population, I think, is sort of an accident. We got lots of urchins last year, this year we have not had a single little baby recruit. I think you are just going to have to control them like you do in the kelp bed. Knock them off. The actual mooring of the plant is going to be a fairly small thing, I am not sure what it will be yet, but it will be a small buoy. There will not be a lot of substrate for urchins.

COON: Are you continuing your fertilization studies in the kelp beds, to assess whether, over the long term, fertilizing is in fact economical?

NORTH: We hope to continue, yes.

CABOT: Have there been any sort of projections on the environmental impact of a whole ocean full of these farms floating miles and miles and bringing massive amounts of deep water to the surface?

NORTH: The first calculations on this possible environmental impact of large scale ocean farms are beginning to come out. It looks like the environmental effect of the greatest potential is the altering of the temperature of the sea surface over a substantial area, because the deep water is usually 10 C or less. If you put a kelp farm in a critical place in the ocean, for example - there is an area in the Indian Ocean, off Africa, that is a very sensitive area, it determines the length and the intensity of the monsoon season over much of southern Asia, -- if you had, I think, between 10 and 20 of these farms sited in that area you could affect the monsoon season. There is also the suggestion that if there is a number of these farms offshore from an upwelling area that they might inhibit the offshore movement of the surface water and reduce the amount of upwelling. If they ever materialize, I think, one would have to control where these farms are sited.

SCHMITZ: You mentioned that there is a very high efficiency of converting the carbon into methane. Did you use specially selected strains of bacteria?

NORTH: Well, these studies are being done at the Institute of Gas Technology in Chicago. They are using a number of mixed cultures from different sources. It turns out the most efficient source was mud that we sent them from the bottom of a submarine canyon, where huge piles of kelp deteriorate. You do not have to heat the water up -- they are not a thermophilic strain. They work at room temberature, which is a big saving in energy.

DRUEHL: Are you paralleling this systems research with selection work on your plants? Are they optimal plants which you put out, once you get your environment set up?

NORTH: Neushul Mariculture has the task of trying to get high yield strains. But that work is just in its very beginning stages.

WHYTE: Would there be any advantage in time release fertilization? Capsulate the ammonium sulfate so you would have a timed release.

NORTH: The Osmocote which is a time release fertilizer worked as well as any with the juvenile plants. I think that the problem with the time release is that it probably will sink to the bottom of the bed where the biomass is not what it is in the upper parts of the water column. I would think that your efficiency might not be as good.

SRIVASTAVA: Is there any possibility of producing gasohol from kelp?

NORTH: The managers at General Electric feel all you have to do is get the right bacteria and they could probably produce ethanol or methanol. The Institute of Gas Research and the Department of Energy are interested in that angle right now. If the concept proves feasible, there will probably be initially food farms and the energy will be a byproduct. Then as the cost of producing energy rises, as our fossil fuels give out, by then the principal product of these oceanic farms may become energy.

MACROCYSTIS HARVEST STRATEGY IN BRITISH COLUMBIA

L.M. Coon

Marine Resources Branch, Ministry of Environment,
Parliament Bldgs., Victoria, British Columbia, Canada V8V 1X5

Introduction

This paper provides an account of resource management research on the giant kelp Macrocystis integrifolia Bory in British Columbia from Scagel's pioneering work (19) in 1947 to the present time. Emphasis is placed on studies aimed at developing an optimal harvest strategy for this valuable renewable resource. Our approach has been to observe how Macrocystis responds to various experimental harvest treatments, and to correlate that information with the seasonality of Macrocystis growth. Inevitably, we have drawn upon the results of the more basic research on Macrocystis biology by other scientists, some of which is reviewed elsewhere in this volume.

Druehl (9) catalogued the known distribution of M. integrifolia in British Columbia and showed that two factors, annual sea surface temperature and salinity, interact to limit the distribution of this kelp (Fig. 1). Estimates of Macrocystis bed area and total standing crop biomass have been made by Foreman (12) and by Marine Resources Branch staff (4,5,7,10,11). Our present estimates indicate a standing crop of 34,600 wet tonnes of M. integrifolia in beds covering a total of 2,260 hectares. This amount represents only 6% of the total standing crop of canopy forming kelps in British Columbia, the remaining 94% being contributed by Nereocystis luetkeana.

An understanding of the complex growth pattern of M. integrifolia was first provided by Scagel (19); this scheme was later modified and elaborated by Lobban (14). Lobban (15) also reviewed the pattern of frond growth and senescence and the reproductive strategy of Macrocystis. The

Synthetic and Degradative Processes in Marine Macrophytes

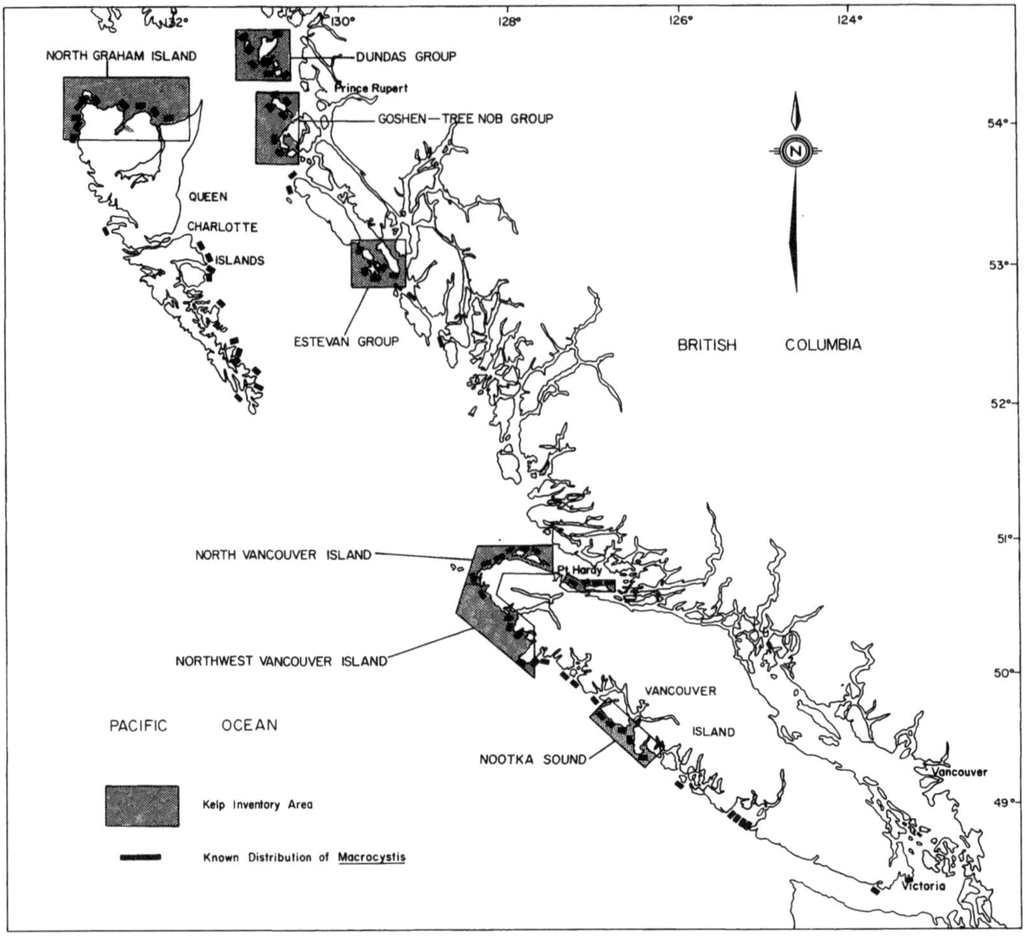

Fig. 1: The distribution of Macrocystis integrifolia in British Columbia (modified after [9]), and the location and extent of areas inventoried using Foreman's (12) KIM-1 technique.

translocation of photoassimilates and translocation patterns in Macrocsytis at various stages of frond development have been studied and are reported elsewhere in this volume (20).

1. The early years of kelp research in British Columbia

Kelp research in British Columbia began with Cameron's (3) inventory of kelp resources over large portions of the British Columbia coastline, which was made in response to concern over North American potash supplies during the First World War. Similarly, Scagel's commissioning by the Provincial Department of Fisheries to study "marine plants of economic importance" in the Port Hardy area (19) was in response to an effort to produce kelp meal and kelp-based cosmetic products in that region.

Scagel (19) measured frond growth rate in M. integrifolia plants having as few as two fronds and on fronds ranging between 11.5 and 442 cm in length. While he noted the great variation in growth rates he did not relate this variance to frond length as was done by later workers (for example, [15]). He found that "major stipes" grew about 5 cm/day; the maximum stipe elongation recorded averaged 7.9 cm/day over a 29 day period. He also observed that removal of the upper 60 cm of frond tissue (i.e., the apical meristem) made it impossible for further production of laminae to take place and subsequent elongation of the stipe was comparatively small (maximum of 0.65 cm/day over a 30 day period).

On the basis of these observations, Scagel (19) proposed the first harvest strategy for M. integrifolia. He focused on season and frequency of harvest and cutting height above the sea bed as the key factors and thought that a shallow cut made early in the growing season would, among other things, thin the canopy, effect a spreading out of the basal meristem, and improve yield on subsequent harvests. This early harvest could, he suggested, be followed by a shallow cut in early summer and a second major and deeper cut in late summer. Scagel also suggested that, since reproductive tissues would not be removed, there was no need to leave parts of the bed unharvested.

2. Macrocystis harvest impact study in the Queen Charlotte Islands

Renewed interest in determining harvesting impacts on Macrocystis resulted from a proposal to initiate a large scale kelp meal operation on the north coast of the Queen Charlotte Islands (see reference 21). Other than estimates of bed area and total standing crop (1,2), virtually nothing was known about Macrocystis in this northern area. Accordingly, a field study (6) was conducted, May - September 1974, to provide information on seasonal growth rates, on the effects of harvesting at two times during the growing season, and on the effect of the position of plants within

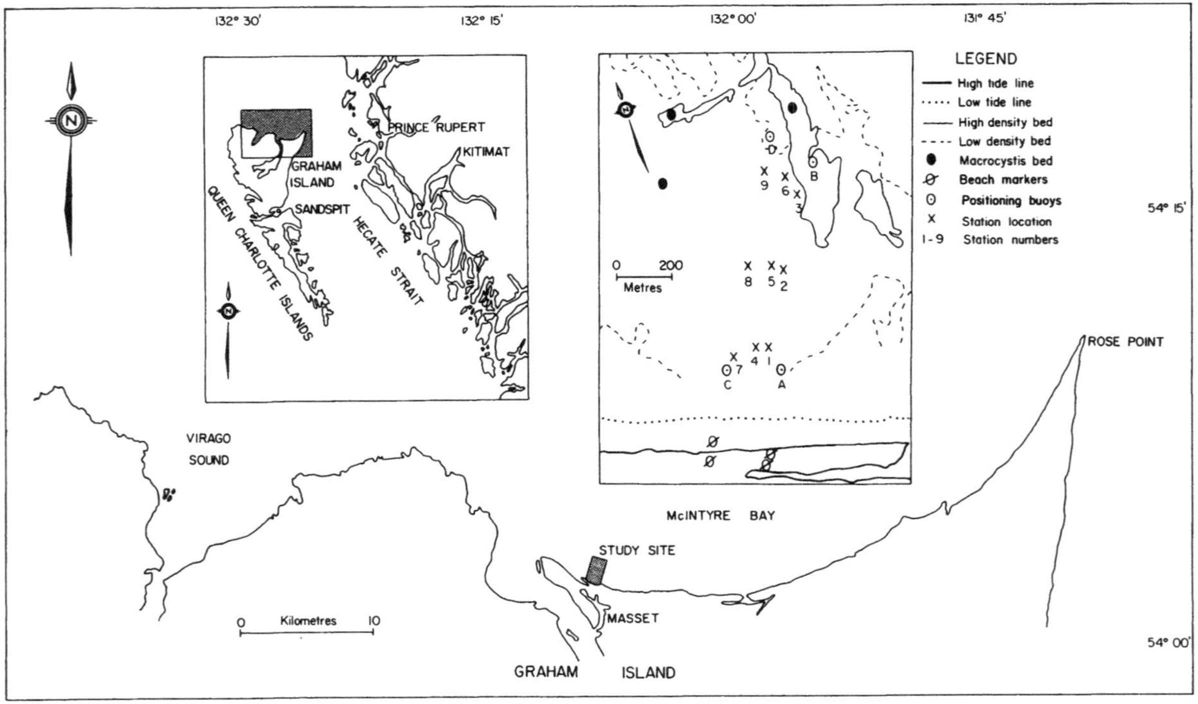

Fig. 2: Location of the Macrocystis harvest study site near Masset, Queen Charlotte Islands in the summer of 1974. The bed was about 1 km wide, of low density (0.33 fronds/m^2) with depth ranging from 0 to -3 m below ZTL (MWL = +2.9 m; tidal range = 6 m). Macrocystis plants were attached to occasional boulders and firm clay in an otherwise sandy substrate. Salinity varied between 31-33 $^o/oo$ and surface temperature between 6-12 C (max. September). The bed is protected from the full force of oceanic surge by several offshore bars.

the bed on frond growth rates.

Three depth zones were delineated within the experimental kelp bed (Fig. 2). Three stations were located in each of these zones, and each station was randomly assigned one of three harvesting treatments: control (unharvested), early harvest (June 7), or late harvest (July 17). At each station, five plants were randomly chosen and tagged. Fronds of tagged plants at harvest stations were cut at 1.6 m below MWL and all nearby plants were similarly cut to prevent shading. Immediately following this cutting, eight fronds with intact apical scimitars were labelled on each of five plants at each station. All tagged fronds were of similar size (mean length for each station varied from 40 to 105 cm). Initially, and at subsequent intervals of 8 to 33 days, lengths of tagged fronds were measured by SCUBA divers using a cloth measuring tape to within $\pm$ 10 cm. The number of blades on each frond and the presence or absence of sporophylls and their condition were noted. Additional short fronds were tagged and measured at the control and early harvest stations on July 26.

Growth rate was calculated as:

$$GR = (L_t - L_0) / t$$

where L_0 was the length of the frond at the beginning of the growth period, L_t was the length at the end of the period, and t was the duration of the period in days. Following North's (18) assumptions, the standard growth rate (G), the percent daily elongation of a 1 m frond, was then calculated. Fronds having a negative growth rate during any of the growth periods, or that were slow growing or did not grow after tagging (< 11% of the fronds at any station), were eliminated from this computation (see [18] for reasons). Standard growth rate (G) is representative of the actively growing, not the total, population of fronds. It is used because the growth rate of fronds is related to frond length (14), and, since it is a singular term which can be used to represent the growth of a whole population of fronds, it serves as a comparative index of growth.

Figure 3 shows the G values for control plants and plants harvested in early June. It is clear from the data on control plants that plants

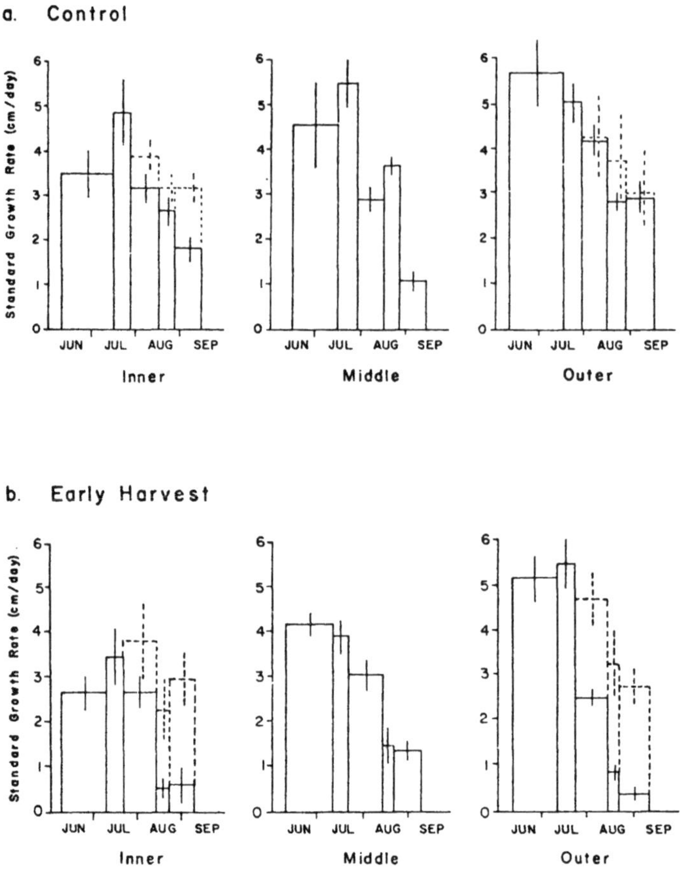

Fig. 3: Mean standard growth rates of <u>Macrocystis</u> fronds at the inner (-0.7 m depth), middle (-1.5 m depth) and outer (-2.3 m depth) stations in the kelp bed between early June and mid September, 1974. (a) control; (b) early harvest at 1.6 m below MWL. Dashed lines indicate new fronds tagged about July 26. Vertical lines at the top of the bars denote 95% confidence interval about the mean.

growing near the outer edge of the bed displayed greater frond elongation rates than those growing near the inner edge, and that the growth rates decreased after July. However, for the shorter fronds tagged on July 26, this decline in late summer was not as marked. The harvested plants showed similar trends and furthermore show that harvesting resulted in reduced frond elongation rates, especially at the inner site, and that this rate did not recover over the rest of the experimental period.

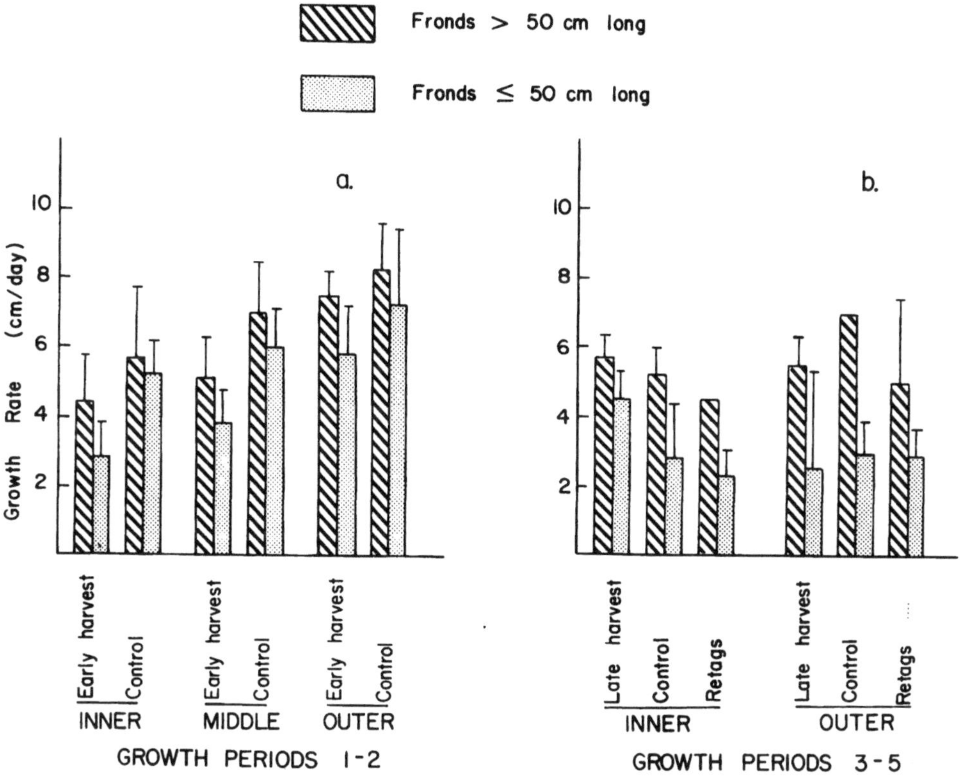

Fig. 4: Frond growth rates (cm/day) in control and harvested plants, (a) early harvest (about June 8); (b) late harvest (about July 15). Control fronds in (b) were tagged on July 28 and their growth was compared with fronds on both late harvest plants and additional fronds (retags) on early harvest plants which were tagged on July 26. Vertical lines intersecting the top of the bars denote the 95% confidence interval about the mean. No confidence interval is given where n<5. Growth periods 1 and 2 spanned the time between June 8 and July 29, while periods 3 to 5 covered from July 29 through September 16, 1974.

Figure 4 compares the mean growth rates of fronds $\leq$ 50 cm and > 50 cm in length in control and harvested plants during early and late summer periods. Data used in this figure represent the entire sample population which lived between consecutive measurement dates; slow growing fronds were not eliminated. Analyses of variance were employed to determine the significance of effects of harvesting, station location or depth of the kelp bed, and initial length of fronds ($\leq$ 50 cm or > 50 cm) on actual growth rates of fronds. Mean growth rates of fronds $\leq$ 50 cm were always significantly less ($P<0.001$) than those recorded for fronds >50 cm for any comparable treatment or time period (see also 6). The effect of station position (or depth) on growth rates was significant ($P<0.001$) following the early harvest, but not after the late harvest. Following the early harvest, mean frond growth rates of plants were lowest at the inner bed stations and greatest at the outer bed stations. Harvesting in early June also resulted in significantly lower ($P<0.001$) growth rates for both size classes of fronds than in comparable controls - the growth rate was reduced by 1 - 2 cm/day throughout the study period. Whereas frond growth was clearly retarded following the early harvest, there was a slight enhancement in growth of harvested plants between early and late harvest at the inner bed locations and this enhancement was particularly evident for the fronds $\leq$ 50 cm.

The G values recorded in the study were slightly higher than those reported for Macrocystis integrifolia plants in the more southerly Barkley Sound, B.C. (3.9 in May, 4.3 in mid-June, and 1.1 in August - October; see refs. 13,14). This may be related to the increased day length experienced by the more northerly plants during summer, but it should be noted that M. pyrifera in southern California has been reported to have G values averaging 7 cm/day over the entire year (18).

The effect of station position (or depth) on growth rate of fronds is difficult to assess. Druehl (8) found that the elongation rate of M. integrifolia fronds on plants transplanted to different depths declined as depth increased. We noted, however, that elongation of fronds on control plants increased from inner bed to outer bed stations. The depth differential was only 2 m, however; and the deeper stations were near the outer edge of the bed where water motion, nutrient availability and light may have been more favourable for Macrocystis growth.

An enhancement of frond growth, following harvest, may be related to increased light availability (18) and/or to a redirection of photoassimilate from the mature blades of the harvested frond to a young frond (instead of being translocated to the apical scimitar of the cut frond, the photoassimilate is now translocated to a different, but young frond, see 13,16 for details). A retardation of growth of juvenile fronds may occur, however, if a sufficient number of mature blades is not left on the cut fronds (13,18). Since, in our study, the relative growth rates of the inner bed plants, following early harvest, were enhanced by mid-July, it seems that the reduced amount of mature tissue remaining on the innermost plants was compensated for by some mechanism, possibly increased light, such that the growth rate of these plants was not reduced relative to that of plants growing further offshore in water 1.4 m deeper. This conjecture is supported by the reduced differential between harvested and control plants at inner and outer bed stations in late summer. Here the implication is that the general reduction in growth rate noted in late summer (Fig. 3) would tend to mask differences due to station position or harvest treatment. It is generally conceded that reduced plant growth in fall in temperate and boreal latitudes is directly related to reduced light availability. These observations then reflect for M. integrifolia the importance of the interplay of light availability and photosynthetic surface area remaining after harvest, which was reported for M. pyrifera (18).

3. Macrocystis harvest strategy trial near Port Hardy, 1976

An existing law in British Columbia stipulates that Macrocystis harvesters must be so constructed that it is impossible to cut fronds deeper than 5 feet (1.5 m) below the water surface. By itself this law will not protect Macrocystis resources from overharvesting because it does not address the problems posed by variable tidal height and the depth range within which Macrocystis grows. For example, the Macrocystis bed near Masset occupies a depth range of 0.8-2.1 m below zero tide level: a harvester cutting at 1.5 m on a zero tide would destroy this bed. An analysis of the results reported by Scagel (19), Lobban (13) and Coon and Roland (6) showed that there might be a direct relationship between the length of frond remnants and the post-harvest elongation rates of juvenile fronds. We attempted to quantify this relationship by making experimental cuts at various heights above the sea bed and measuring subsequent biomass recovery over time. We expected that both season of harvest and frequency of harvest would modify the effect of cutting height; both factors were therefore integrated into the experimental design. This approach offers several practical advantages as a basis for a harvest strategy:

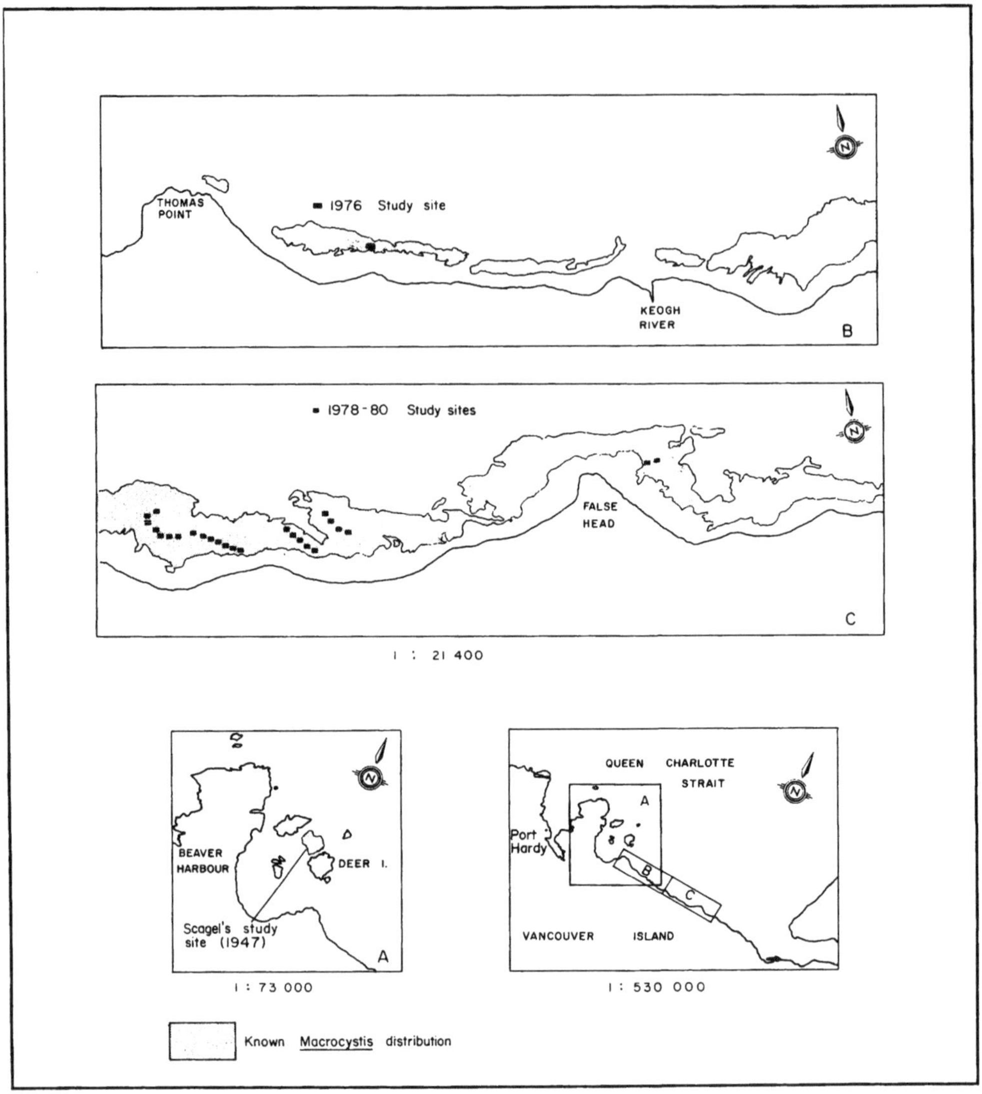

Fig. 5: Location of Macrocystis study sites in the Port Hardy region: (A) Scagel's (19) Deer Island site, (B) the 1976 site used by Coon and Lobban, (C) the current harvest strategy site showing positions of the 25 treatment plots, 40 m x 30 m each. Macrocystis plants in (B) and (C) are attached to cobbles and boulders imbedded in a mixed shell-sand substrate and to occasional low relief bed rock outcroppings. Mean monthly surface temperature ranges from 8.5 to 12.5 C and salinity from 30.7 to 31.5 $^{o}/oo$ (ref. 9) in this area. The 1976 site is in a medium density bed (0.95 fronds/m^2) about 180 m wide and occupies a depth range of -1.5 to -4.5 m. The current site is also in a medium density bed (0.85 fronds/m^2) with a mean depth between -2.5 and -3.5 m.

a. biomass recovery after various harvest treatments can be relatively easily monitored by mechanical harvesting and provides a quantitative evaluation of the general condition of plants and their overall net response to each treatment;

b. once a minimum cutting height is established, the harvester need only employ a depth sounding device to ensure that the cutting blade is kept the appropriate distance above the sea bed; and

c. enforcement of harvesting regulations based on cutting height may be simplified because a harvester would leave behind a conservative record of his cutting activities in the form of cut stipes which persist for at least several weeks.

The Macrocystis bed used in the study lies off the Port Hardy airport (Fig. 5). Divers laid down and anchored a premeasured polypropylene rope grid outlining eighteen 10 m x 10 m quadrats in the mid-portion of the selected site (site B, Fig. 5). Bottom depth within the grid averaged -2.5 m (range -1.85 to -3.25 m). The grid was divided into two blocks of 9 quadrats each. The northwestern block was harvested first on June 10-11, again on August 14-15, and again on September 28-29, 1976; the southeastern block was also harvested on June 10-11, but was not reharvested until September 28-29, 1976. Each block contained 3 replicate quadrats for each of 3 cutting height treatments: 1.5, 3.0, and 4.5 m above the sea bed. The 1.5 m cutting height approximated the deepest possible cut which could be made by a commercial harvester operating during a low tide period. No effort was made to randomize the placement of either treatments or replicates within each block. Divers hand-harvested all Macrocystis fronds within each quadrat at the appropriate height above the sea bed. Harvested fronds were brought to a boat, stored in labelled plastic containers and transported to Port Hardy where wet weights were recorded using a scale accurate to $\pm$ 0.045 kg.

Table I summarizes the initial and subsequent yields obtained.

An analysis of variance of initial yields (n=6) for each treatment showed that variance within treatments masked differences between treatments. This resulted from the great spatial heterogeneity of Macrocystis plants within this bed. In order to make comparisons between treatments it was

Table I: Absolute yield and percent regrowth of M. integrifolia as affected by cutting height, season and frequency of harvest in 1976. Yield figures are totals of 3 replicate plots; each plot was 100 m^2 in area.

Harvest Schedule	Cutting Height Above Seabed (m)	Initial Harvest Yield (wet kg)	1st Reharvest Yield (wet kg)	2nd Reharvest Yield (wet kg)	Combined Reharvest Yield (wet kg)	Total Yield per Treat. (wet kg)
June-Sep[a]	1.5	657	116 (18)[b]			773
June-Sep	3.0	556	171 (31)			727
June-Sep	4.5	517	190 (37)			707
June-Aug-Sep[c]	1.5	921	218 (24)	45 (5)	263 (29)	1184
June-Aug-Sep	3.0	619	315 (51)	46 (7)	361 (58)	980
June-Aug-Sep	4.5	323	218 (67)	75 (23)	293 (91)	616

a Data from southeastern block.
b Data in parentheses are % of initial yield.
c Data from northwestern block.

necessary to express the reharvest yields for each treatment (n=3) as a percentage of the initial harvest yield for that treatment. Thus all statistical analyses reported herein are based on data which have been normalized by angular transformation. A second analysis of variance was then conducted to test for significant differences in relative regrowth between treatments for each regrowth period. A Student-Newman-Keuls test was performed to determine which treatment means were significantly different from the others.

Foreman (12) had noted that the distribution of Macrocystis biomass in the water column is essentially linear. Accordingly, we were not surprised to find a decrease in initial yield as cutting height increased; this pattern can also be noted in the total yields but results largely from the yield differences obtained from the initial harvest. While the yields for the mid-August reharvest for the 1.5 m and 4.5 m treatments are the same (218 kg), they represent quite different percent regrowths. In fact percent regrowth increased significantly with each cutting height increment in the 2 month period following the initial harvest. In the 2 month period following the mid-August reharvest, however, the percent regrowths for the 1.5 m and 3.0 m treatments were very similar and both were significantly lower than that of the 4.5 m treatment. It is noteworthy that at the end of September the 4.5 m treatment plots contained plants of normal, vigorous appearance, while the plants in the 1.5 m plots were in poor condition, exhibiting fewer fronds and no canopy.

While there was a definite increase in yield and percent regrowth with increased cutting height in the plots reharvested after a 4 month interval, the differences between the percent yields of the three treatments were not significant. The low biomass recovered after the 4 month regrowth period is indicative of the action of a seasonal factor which tended to reduce the overall Macrocystis biomass in the last half of August and September.

This investigation showed that total harvest yield declined as cutting height increased, even if two or three harvests are made. Percent regrowth, however, increased with cutting height. The effects of the onset of autumn on regrowth potential tended to mask those related to cutting height. On a more qualitative basis it was observed that general plant condition improved with increased cutting heights. Finally we determined that a quadrat size of 100 m^2 was inadequate due to the highly variable density of Macrocystis plants in our study area. Our options for later studies were to increase the number of replicates for each treatment and/or to increase the quadrat size.

4. The Macrocystis harvest strategy program (1978-)

An expanded harvest program was designed to evaluate the effects of cutting heights, season of harvest, and interval between harvests on percent regrowth and yield (Table II). The once yearly harvest treatments, aside from illustrating the interactive effects of season of cut and cutting height on regrowth, served as standards against which the impact of two and three harvests per year was measured. The twice yearly harvest treatments represent what we expect, in an a priori sense (see 19), to be within the realm of commercial potential. The thrice yearly harvest treatments provide experimental "worst case" situations in which

Table II: Experimental design and harvest schedule employed in the Macrocystis harvest strategy program. Initial harvests were made in 1979. Plots were harvested at each of three depths (1.5, 3.0 and 4.5 m) above the seabed, as noted in parentheses. These plots were harvested at the times indicated by asterisk. Plot size was 1200 m^2.

Harvesting Frequency	May	June	Jul	Aug	Sep	Harvests in 1980 - 1981
One harvest per season	* (3)					May
		* (3)				June
			* (3)			July
				* (3)		August
					* (3)	September
Two harvests per season at 3 month intervals	* (3)			*		May-Aug
		* (3)			*	June-Sept
Three harvests per season at 2 month intervals	* (3)		*		*	May-Jul-Sep

we monitored harvest impacts on individual Macrocystis plants, understory macrophytes, benthic macroinvertebrates and detritus standing crop. The three plots receiving this most severe harvest treatment, together with the control plot are called the "monitoring plots." In the following, I summarize only the harvest yields in the first year of the experiment. Harvesting impacts in the monitoring plots are still to be analyzed. The details of the study site near Port Hardy and field methods are given in figure 5.

We had hoped that increasing plot size to 1200 m^2 would overcome the effects of spatial heterogeneity of Macrocystis biomass noted in the preceding study. The initial yields from plots receiving the same cutting height treatment were highly variable (cf. for instance, the May yields at 1.5 m or at 3.0 m, Table III). An analysis of variance of the May data illustrated that even with 3 replicates, the variance within treatments was sufficient to mask the variance between treatments. As a result we have reverted to the use of percent regrowth data, as defined in the preceding study, in order to compare treatment effects.

Table III: Absolute yield and percent regrowth of M. integrifolia as affected by cutting height, season and frequency of harvest in 1979. Plot size was 1200 m^2.

Harvest Schedule	Cutting Height Above Seabed (m)	Initial Harvest Yield (wet kg)	1st Reharvest Yield (wet kg)	2nd Reharvest Yield (wet kg)	Combined Reharvest Yield (wet kg)	Total Yield per Treat. (wet kg)
May	1.5	1924				1924
May	3.0	3636				3636
May	4.5	2078				2078
June	1.5	2067				2067
June	3.0	2694				2694
June	4.5	2130				2130
Jul	1.5	3828				3828
Jul	3.0	3286				3286
Jul	4.5	1358				1358
Aug	1.5	7808				7808
Aug	3.0	3314				3314
Aug	4.5	1185				1185
Sep	1.5	2194				2194
Sep	3.0	1935				1935
Sep	4.5	1716				1716
May-Aug	1.5	3798	810 (21)[a]			4608
May-Aug	3.0	1534	1225 (80)			2759
May-Aug	4.5	1351	1810 (134)			3161
Jun-Sep	1.5	3602	788 (22)			4399
Jun-Sep	3.0	2290	784 (34)			3074
Jun-Sep	4.5	2547	2351 (92)			4898
May-Jul-Sep	1.5	3485	585 (17)	245 (7)	830 (24)	4315
May-Jul-Sep	3.0	2400	862 (36)	390 (16)	1252 (52)	3652
May-Jul-Sep	4.5	1309	1231 (94)	602 (46)	1833 (140)	3142

[a] Data in parentheses are % of initial yield.

Table III provides the initial harvest and, where applicable, subsequent reharvest yields in absolute terms and as percentages of the initial yields for each treatment. For the twice harvested plots the lowest percent regrowth was recorded in those cut at 1.5 m, and the highest in plots cut at 4.5 m. While there was little difference in percent regrowth between the May - August and the June - September plots for the 1.5 m treatment, reductions of 46 and 42% were recorded for the 3.0 m and 4.5 m treatments, respectively, between these same harvest schedules. Such reduced regrowth resulting from delaying harvesting and reharvesting by only one month may be related both to: a) increased growth rates in May relative to June, as has been recorded for other Laminariales in temperate waters (see ref. 17, for example), and b) a decline in growth rates between August and September as has been noted for M. integrifolia at several sites in British Columbia (6,15).

For the plots harvested three times in one season, most of the percent regrowth at all cutting heights, occurred in the interval between the initial cutting in May and the first reharvest in July. This is probably related to the seasonally normal rapid growth of M. integrifolia in spring-early summer. The combined yield (in both weight and %) from both reharvests was not greatly different from that observed in the May - August harvest plots, indicating little practical value of making three harvests per year.

In all cases the absolute yields of the initial harvests were greatest for the 1.5 m treatment plots and least for the 4.5 m plots, as expected given the linear vertical distribution of Macrocystis biomass (12). However, the absolute yields upon reharvesting in all cases reverse this trend. This has a tendency to reduce the differences in absolute yield between cutting height treatments.

At this point in time it would appear that restricting Macrocystis harvesting to no closer than 4.5 m above the sea bed would optimize percent regrowth and lead to canopy replenishment in the shortest possible time without overly constraining absolute yield - particularly if two or three harvests per year are to be attempted. We plan on continuing this experimental series through 1980 and 1981 to see how Macrocystis responds to harvesting over a three year period.

Acknowledgements

The 1976 study in the Port Hardy area was conducted in collaboration with Dr. C.S. Lobban to whom grateful thanks are extended.

References

1. Anon.: Survey of the kelp resource within 18 miles of Masset. North Pacific Marine Products Ltd., 52 pp. 1967

2. Blakely, B.B., Chalmers, W.T.: Masset kelp inventory. Final report to Dept. Environment, Fisheries Operations, 72 pp. 1973

3. Cameron, A.T.: The commercial value of the kelp beds of the Canadian Pacific Coast - A preliminary report and survey of the beds. Contrib. Can. Biol., 1914-1915, sessional paper #38a, pp. 24-39 1916

4. Coon, L.M., Field, E.J., Canadian Benthic Ltd.: Nootka Sound kelp inventory, 1975. Fisheries Management Report #2, British Columbia Marine Resources Branch 27 pp. 1976

5. Coon, L.M., Roland, W., Field, E.J., Clayton, W.E.L.: Kelp inventory, 1976. Part III. North and west coasts Graham Island (Queen Charlotte Islands). Fisheries Development Report #13, British Columbia Marine Resources Branch. 26 pp. 1979

6. Coon, L.M., Roland, W.G.: Harvesting impacts on *Macrocystis integrifolia*: a preliminary study. Fisheries Development Report #12, British Columbia Marine Resources Branch, 46 pp. 1980

7. Coon, L.M., Roland, W.G., Field, E.J., Clayton, W.E.L., Jenson, V.: Kelp inventory, 1976: Part IV. Goschen Island to the Tree Knob Group. Fisheries Development Report #19, British Columbia Marine Resources Branch, 19 pp., 1980

8. Druehl, L.D.: Proc. Int. Seaweed Symp. 9, 79-84 (1978).

9. Druehl, L.D.: Can. J. Bot. 56, 69-79 (1978).

10. Field, E.J., Coon, L.M., Clayton, W.E.L., Clark, E.A.C.: Kelp inventory, 1976. Part I. The Estevan Group and Campania Island. Fisheries Management Report #9, British Columbia Marine Resources Branch, 19 pp. 1977

11. Field, E.J., Clark, E.A.C.: Kelp inventory, 1976. Part II. The Dundas Group. Fisheries Management Report #11, British Columbia Marine Resources Branch, 24 pp. 1978

12. Foreman, R.E.: KIM-1. A method for inventory of floating kelps and its application to selected areas of Kelp Licence Area 12. Benthic Ecological Research Program Report 75-1. Report to Federal Fisheries and Marine Service and Provincial Marine Resources Branch, 81 pp. (ms.) 1975

13. Lobban, C.S.: Growth, translocation and harvesting interactions in Macrocystis integrifolia. Report to British Columbia Marine Resources Branch, Ministry of Environment, 8 pp. (ms.) 1976

14. Lobban, C.S.: Can. J. Bot. 56, 2707-2711 (1978).

15. Lobban, C.S.: Phycologia 17, 196-212 (1978).

16. Lobban, C.S.: J. Phycol. 14, 178-182 (1978).

17. Mann, K.H., Chapman, A.R.O., Gagne, J.: In Synthetic and Degradative Processes in Marine Macrophytes. (L.M. Srivastava, ed.), Walter de Gruyter, Berlin, 1981

18. North, W.J.: Nova Hedwigia 32, 75-92 (1971).

19. Scagel, R.F.: An investigation on marine plants near Hardy Bay, B.C. Report to the Provincial Department of Fisheries, #1, 70 pp. 1947

20. Schmitz, K.: In Synthetic and Degradative Aspects of Marine Macrophytes (L.M. Srivastava, ed.) Walter De Gruyter, Berlin 1981

21. Whyte, J.N.C., Englar, J.R.: Fish. Res. Board Canada, Tech. Rep. 453 (1974).

SUBJECT INDEX

Bei Fragen zur Produktsicherheit wenden Sie sich bitte an:
If you have any questions regarding product safety,
please contact:

Walter de Gruyter GmbH
Genthiner Straße 13
10785 Berlin
productsafety@degruyterbrill.com